aus der Reihe:

Innovationen mit Mikrowellen und Licht

Forschungsberichte aus dem Ferdinand-Braun-Institut, Leibniz-Institut für Höchstfrequenztechnik

Band 66

Sebastian Walde

AlN base layers for UV LEDs

Herausgeber: Prof. Dr. Günther Tränkle

Ferdinand-Braun-Institut
Leibniz-Institut
für Höchstfrequenztechnik (FBH)
Gustav-Kirchhoff-Straße 4
12489 Berlin

Tel. +49.30.6392-2600
Fax +49.30.6392-2602

E-Mail fbh@fbh-berlin.de
Web www.fbh-berlin.de

Innovations with Microwaves and Light

Research Reports from the Ferdinand-Braun-Institut, Leibniz-Institut für Höchstfrequenztechnik

Preface of the Editor

Research-based ideas, developments, and concepts are the basis of scientific progress and competitiveness, expanding human knowledge and being expressed technologically as inventions. The resulting innovative products and services eventually find their way into public life.

Accordingly, the *"Research Reports from the Ferdinand-Braun-Institut, Leibniz-Institut für Höchstfrequenztechnik"* series compile the institute's latest research and developments. We would like to make our results broadly accessible and to stimulate further discussions, not least to enable as many of our developments as possible to enhance everyday life.

The author investigates approaches for AlN base layers to improve the efficiency of UV LEDs deposited on them. This is accomplished by reducing the threading dislocation density through high-temperature annealing, which leads to increased internal efficiency. Nanostructuring of the surface of the sapphire substrate on which the AlN layer is deposited helps improving the light extraction. For this approach an intermediate high-temperature treatment step is necessary to control the strain of the layer stack and to avoid cracks. This work lays the foundation for further improvement of UV LEDs as needed for disinfection of water and surfaces.

We wish you an informative and inspiring reading

Prof. Dr. Günther Tränkle
Scientific Director

The Ferdinand-Braun-Institut

The Ferdinand-Braun-Institut researches electronic and optical components, modules and systems based on compound semiconductors. These devices are key enablers that address the needs of today's society in fields like communications, energy, health and mobility. Specifically, FBH develops light sources from the visible to the ultra-violet spectral range: high-power diode lasers with excellent beam quality, UV light sources and hybrid laser systems. Applications range from medical technology, high-precision metrology and sensors to optical communications in space and integrated quantum technology. In the field of microwaves, FBH develops high-efficiency multi-functional power amplifiers and millimeter wave frontends targeting energy-efficient mobile communications as well as car safety systems. In addition, compact atmospheric microwave plasma sources that operate with economic low-voltage drivers are fabricated for use in a variety of applications, such as the treatment of skin diseases.

The FBH is a competence center for III-V compound semiconductors and has a strong international reputation. FBH competence covers the full range of capabilities, from design to fabrication to device characterization.

In close cooperation with industry, its research results lead to cutting-edge products. The institute also successfully turns innovative product ideas into spin-off companies. Thus, working in strategic partnerships with industry, FBH assures Germany's technological excellence in microwave and optoelectronic research.

AlN base layers for UV LEDs

vorgelegt von
Master of Science in Physik

Sebastian Walde

von der Fakultät II - Mathematik und Naturwissenschaften
der Technischen Universität Berlin
zur Erlangung des akademischen Grades

Doktor der Naturwissenschaften
- Dr. rer. nat. -

genehmigte Dissertation

Promotionsausschuss:

Vorsitzende:	Prof. Dr. Kathy Lüdge
Gutachter:	Prof. Dr. Michael Kneissl
Gutachter:	Prof. Dr. Ferdinand Scholz
Gutachter:	Prof. Dr. Markus Weyers

Tag der wissenschaftlichen Aussprache: 15.12.2020

Berlin 2021

Bibliografische Information der Deutschen Nationalbibliothek
Die Deutsche Nationalbibliothek verzeichnet diese Publikation in der Deutschen Nationalbibliografie; detaillierte bibliografische Daten sind im Internet über http://dnb.d-nb.de abrufbar.

1. Aufl. - Göttingen: Cuvillier, 2021
Zugl.: (TU) Berlin, Univ., Diss., 2021

Nonnenstieg 8, 37075 Göttingen
Telefon: 0551-54724-0
Telefax: 0551-54724-21
www.cuvillier.de

1. Auflage, 2021
Gedruckt auf umweltfreundlichem, säurefreiem Papier aus nachhaltiger Forstwirtschaft.

ISBN 978-3-7369-7451-7
eISBN 978-3-7369-6451-8

Abstract

In this work, the aim is to improve the quality of AlN base layers on sapphire substrates to enable the fabrication of high performance ultraviolet (UV) light-emitting diodes (UV LEDs). The main issues for UV LEDs are still a limited internal quantum efficiency due to a high amount of threading dislocations, but also a limited light extraction efficiency due to total internal reflection at the AlN/sapphire interface. The recently emerging methods of high temperature annealing of AlN/sapphire layers and growth on nanopatterned sapphire substrates are potential candidates for solving those issues and hence, are comprehensively investigated in this work.

High temperature annealing was applied to AlN layers of different strain and thickness grown by metalorganic vapour phase epitaxy (MOVPE). For a 900 nm thick AlN layer annealed at 1700 °C for 3 h, the threading dislocation density was successfully reduced by more than one order of magnitude down to $6 \times 10^{8}\,\mathrm{cm}^{-2}$. Extended analysis revealed that the in-plane strain of the AlN layers is changed to more compressive values due to high temperature annealing. This has to be considered for subsequent UV LED heterostructure growth. Oxygen diffusion from the sapphire into the AlN during HTA resulted in the formation of AlON in the AlN layer. Nevertheless, the UV transparency was retained.

To analyse the benefit in terms of light extraction efficiency by using nanopatterned sapphire substrates, wave optical simulations of on top grown UV LEDs were conducted. For a nanohole pattern with pitch of 1000 nm and an assumed effective reflectivity of the p-side of the UV LED of 0.8, an increase in the transmittance through the AlN/sapphire interface by 77 % compared with a planar interface was calculated. The MOVPE growth process on sapphire nanopillars and sapphire nanoholes was optimised and succeeded in ending up with an atomically smooth surface. For growth on the nanopillars this was reached by reducing the sapphire offcut angle to suppress the formation of step bunches. For growth on the nanoholes an intermediate high temperature annealing step in order to change the strain state to more compressive values enabled coalescence without layer cracking. The threading dislocation density was reduced to approx. $1 \times 10^{9}\,\mathrm{cm}^{-2}$ for AlN on both the nanopillars and nanoholes.

In collaboration with the group of Michael Kneissl at the Technical University of Berlin, UVC LEDs of 265 nm wavelength were grown on top of the developed templates to evaluate their performance. Electroluminescence was achieved for all templates with considerable improvement demonstrated by specific comparisons to other templates.

By comparing two templates with similar threading dislocation density but one with a planar and the other one with a nanopatterned AlN/sapphire interface, the predictions of the wave optical simulations were approved by the experimental results.

Kurzfassung

Das Ziel dieser Arbeit ist die Qualitätssteigerung von AlN-Basisschichten auf Saphirsubstraten, um die Herstellung von leistungsfähigen ultravioletten (UV) lichtemittierenden Dioden (UV-LEDs) zu ermöglichen. Die Hauptprobleme der UV-LEDs sind nach wie vor eine verminderte interne Quanteneffizienz aufgrund hoher Versetzungsdichte und eine verminderte Lichtextraktionseffizienz aufgrund Totalreflektion an der AlN/Saphir-Grenzfläche. Die kürzlich aufgekommenen Methoden des Hochtemperaturerhitzens von AlN/Saphir-Schichten und das Wachstum auf nanostrukturiertem Saphir bieten Lösungen für diese Probleme und werden deshalb in dieser Arbeit umfassend untersucht.

Hochtemperaturerhitzen wurde an AlN-Schichten mit unterschiedlicher Verspannung und unterschiedlicher Schichtdicke durchgeführt, welche mittels metallorganischer Gasphasenepitaxie (MOVPE) hergestellt wurden. Für eine 900 nm dicke Schicht, erhitzt bei 1700 °C für 3 h, konnte die Dichte der durchstoßenden Versetzungen um mehr als eine Größenordnung auf $6 \times 10^8\,\mathrm{cm}^{-2}$ reduziert werden. Durch eine erweiterte Analyse wurde eine Änderung der Verspannung aufgrund von HTA zu mehr kompressiven Werten nachgewiesen, was für das weitere UV-LED-Wachstum berücksichtigt werden muss. Sauerstoffdiffusion vom Saphir in das AlN führte zur Bildung von AlON in der AlN-Schicht. Trotzdem konnte die Transparenz im UV erhalten werden.

Um den Vorteil von der Nutzung nanostrukturierter Saphirsubstrate bezogen auf die Lichtextraktionseffizienz zu untersuchen, wurden wellenoptische Simulationen an darauf gewachsenen UV-LEDs durchgeführt. Für eine Nanostrukturierung mit einem Lochabstand von 1000 nm und einer angenommenen effektiven Reflektivität der p-Seite der UV-LED von 0.8 wurde eine Erhöhung der Transmission durch die AlN/Saphir-Grenzfläche um 77 % im Vergleich zu einer planaren Grenzfläche berechnet. MOVPE-Wachstum auf Saphir mit Nanosäulen und mit Nanolöchern wurde optimiert, sodass eine atomar glatte Oberfläche erreicht werden konnte. Beim Wachstum auf den Nanosäulen wurde dies durch Reduzierung des Substratfehlschnittes und daraus folgender Vermeidung des Aufbaus von Stufenbündeln bewirkt. Beim Wachstum auf den Nanolöchern ermöglichte ein intermediärer Hochtemperaturschritt, der in einer Veränderung der Verspannung zu kompressiven Werten resultierte, Koaleszenz der Schicht ohne Rissbildung. Für das AlN auf den Nanosäulen und den Nanolöchern konnte die Dichte der durchstoßenden Versetzungen auf jeweils ca. $1 \times 10^9\,\mathrm{cm}^{-2}$ reduziert werden.

In Zusammenarbeit mit der Gruppe von Prof. Michael Kneissl an der technischen Universität Berlin wurden UVC-LEDs mit 265 nm Wellenlänge auf die entwickelten

Templates abgeschieden, um ihre Leistungsfähigkeit zu überprüfen. Elektrolumineszenz wurde für alle Templates erreicht mit beachtlicher Verbesserung, was durch spezielle Vergleiche mit anderen Templates demonstriert wurde. Darüber hinaus konnte durch den Vergleich zweier Templates mit ähnlicher Dichte der durchstoßenden Versetzungen, aber mit planarer und nanostrukturierter AlN/Saphir-Grenzfläche, die Vorhersage der wellenoptischen Simulationen experimentell bestätigt werden.

Contents

Introduction and Motivation

The first question that is often asked to a scientist about his research topic from people outside the research community is: "What can it be used for?". For the present work the answer already lies in the title. The topic of interest, the AlN base layers, are used as the foundation for ultraviolet (UV) light-emitting diodes (LEDs). Understandably, the following question will most likely be what UV LEDs are used for?

LEDs in general have a long history of revolutionising illumination technology. While the first super bright red, yellow and green LEDs on basis of GaAsP were developed already in the early 1980s, the real breakthrough of LEDs was achieved by the realisation of the first GaN-based blue LEDs in the late 1980s [1]. The blue light finally enabled the generation of white light by LEDs and thus, accessing their application for standard illumination technology [2]. For this breakthrough Isamu Akasaki, Hiroshi Amano and Shuji Nakamura were awarded with the Nobel prize in physics in 2014 [3]. They managed to fix the problem of p-type doping in GaN which had been the unsolved step towards realising GaN-based blue LEDs for years [4, 5]. There are many advantages of LEDs over conventional light sources, e.g. the potential of being more efficient, the small size making LEDs easy to implement, the upscalable manufacturing process for mass production at low cost, etc. All these advantages will also hold for UV LEDs. To push the wavelength from the blue LEDs down to the UVA (400 nm-320 nm), UVB (320 nm-280 nm) and further to the UVC (280 nm-200 nm) spectral range, Al(In,Ga)N-based heterostructures with underlying AlN base layers on sapphire are used. With such structures it is possible to cover a wavelength range between 210 nm and 340 nm. Since UV light has numerous applications in various fields there is a big motivation to open up this spectral range for LED lighting. Up to now, UV light is produced by low and medium pressure mercury lamps. Apart from the high environmental toxicity of mercury, a high voltage and a relatively long warm-up time is needed to operate mercury lamps. However, the main disadvantage might be the impossibility of choosing the exact wavelength in the UV spectral range, since it is restricted to the characteristic mercury energy transitions. Instead, for AlGaN-based UV LEDs it is possible to specify

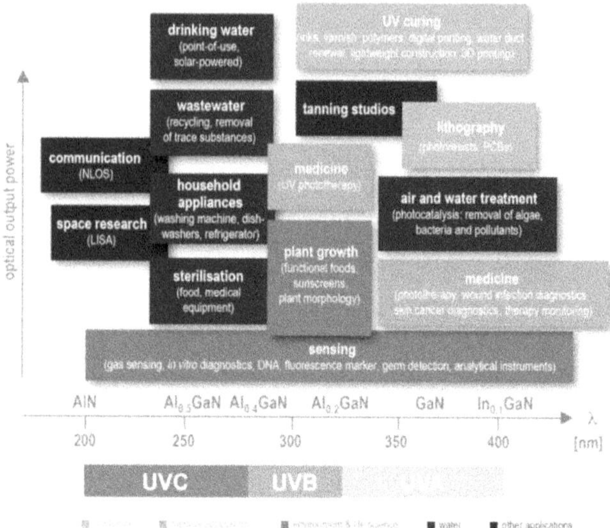

Figure 0.1: Applications of UV LEDs visualised in dependency of the wavelength and the optical power. Additionally, the related AlGaN composition with respect to the wavelength range is shown. From reference [10] with permission of Advanced UV for Life © 2020.

the wavelength by adapting the material composition inside the quantum wells in the active region. Hence, UV LEDs do not only have the potential of replacing mercury lamps regarding already established applications of UV light, but they will also enable new applications for which very specific wavelengths in the UV spectral range are required [6]. An example for such an application is the field of gas sensing, since the wavelength has to fit an absorption band of the specific gas [7–9]. An overview of the different fields of applications depending on the wavelength but also the output power is shown in Fig. 0.1. One of the main application fields is water purification [11–14], but also disinfection and sterilisation [15, 16]. The mechanism behind these applications is that UVC light around 265 nm triggers chemical modification of DNA (or RNA) of microorganisms resulting in the loss of the ability to replicate. Considering the ongoing global COVID-19 pandemic the interest for disinfection has recently increased heavily. Studies of inactivating related coronaviruses by UV light from the 2000s are promising for the hypothesis of UV light being able to also inactivate the coronavirus responsible for the COVID-19 pandemic [17–19]. First research results by Inagaki et al., which isolated SARS-CoV-2 viruses from the cruise ship "Diamond Princess", confirm this for a wavelength of 280 nm [20]. Besides the inactivation of microorganisms UV light in the UVA and UVB range can be used to improve plant growth. Specific UVB radiation for instance leads to the production of beneficial secondary metabolites [21]. For the emerging field of urban indoor farming the whole spectrum of the sun also including UVA and UVB light is relevant making LEDs the best applicable technology [22]. Furthermore, UV LEDs can be used in medicine for phototherapy of skin diseases and medical diagnostics, but also in industry, e.g. for UV curing of polymers [23–27].

The present thesis introduces the basics prior to the main part of the scientific results. The physical basics will be given in the first "Fundamentals" chapter, while technical details of all experimental methods follow in the second "Experimental methods" chapter. The main part of the scientific results begins with chapter 3 "High temperature annealing of MOVPE grown AlN templates" by a comprehensive study of the high temperature annealing (HTA) of AlN base layers grown on sapphire by metalorganic vapour phase epitaxy (MOVPE). The aim of this annealing is the improvement of the material quality. Chapter 4 "Nanopatterned sapphire substrates" deals with AlN growth on nanopatterned sapphire substrates (NPSS) which has the potential to improve not only the material quality of the AlN but also the light extraction efficiency of UV LED heterostructures grown on top. In chapter 5 "UVC LED performance on the developed AlN templates" the outcome of the newly developed preparation techniques for the AlN base layers on sapphire will be evaluated in terms of performance of UVC LED heterostructures grown on top with a targeted wavelength of 265 nm. Each of the main result chapters closes with a short summary, while the whole work ends with a summary of the most important findings and an outlook to the near future of preparing AlN base layers for UV LEDs. This work was done at the Ferdinand-Braun-Institut, Leibniz-Institut für Höchstfrequenztechnik (FBH), in the group of Prof. Markus Weyers.

CHAPTER 1

Fundamentals

In this chapter the basic physical background will be given for the preparation of the AlN base layers including the crystal structure of AlN and the heteroepitaxial relation between AlN and sapphire. Further, the impact of specific properties of the AlN base layers on the performance of UV LED heterostructures grown on top will be discussed with the aim to define certain quality criteria for the AlN base layers regarding the suitability of being used as UV LED templates. Additionally, the state of research in the field of AlN base layers will be outlined to put this work in the context of related literature.

1.1 Heteroepitaxial growth of AlN on sapphire

Prior to describing the heteroepitaxial growth of AlN on sapphire, an answer to the intuitive question will be given why AlN is grown heteroepitaxially on sapphire and why not native AlN bulk substrates are used as base layers similar to other III-V semiconductor material systems, e.g. based on GaAs [28]. Also for the III-nitride material GaN the usage of GaN bulk substrates becomes increasingly popular [29]. However, in case of AlN still many challenges in the complex process of growing AlN bulk substrates have to be overcome. AlN bulk substrates are usually grown by physical vapour transport at temperatures above 2000 °C [30]. In the past a lot of progress has been made in the preparation of AlN bulk substrates in terms of wafer size and crystal quality [31, 32]. However, one of the main issue is still the lack of transparency in the relevant UV range which is ascribed to impurities built in during growth [33–35]. The heteroepitaxial growth of AlN on sapphire instead is a straightforward technology to

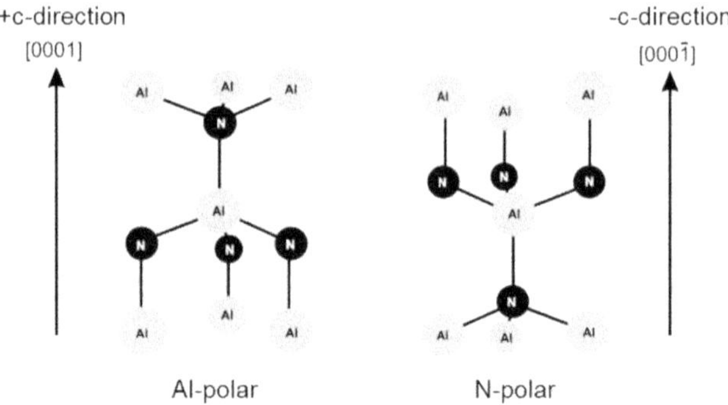

Figure 1.1: Schematic structure of wurtzite AlN in $+c$ and $-c$ direction resulting in Al- and N-polarity, respectively. From reference [42] with permission of Cuvillier © 2014.

go for, since it is similar to the heteroepitaxial growth of GaN on sapphire which is well established in the industrial manufacturing process of GaN-based blue LEDs [36]. Despite the lattice mismatch, GaN and also AlN grow reasonably well on sapphire while sapphire substrates themselves are available in large sizes at low cost [37].

AlN usually has the hexagonal wurtzite crystal structure shown in Fig. 1.1 with lattice constants $a_{\mathrm{AlN}} = 0.3111\,\mathrm{nm}$ and $c_{\mathrm{AlN}} = 0.4981\,\mathrm{nm}$ [38]. The band gap of AlN is relatively high with 6.1 eV [39]. In Fig. 1.1 it can be seen that the $+c$ and $-c$ directions lead to a different arrangement of the Al and N atoms to each other determining the polarity [40]. By definition, in Fig. 1.1 the $+c$ direction is Al-polar, while the $-c$ arrangement is N-polar. Since AlN of Al-polarity or N-polarity show different properties, it is crucial to control the polarity [41]. In this work the desired phase is Al-polarity. The sapphire substrates (α-Al_2O_3) have a trigonal corundum crystal structure with lattice constants $a_{\mathrm{Al_2O_3}} = 0.4758\,\mathrm{nm}$ and $c_{\mathrm{Al_2O_3}} = 1.298\,\mathrm{nm}$ [43]. In this work AlN is grown on (0001) sapphire. Similar to GaN on (0001) sapphire, AlN will grow on the (0001) sapphire surface with (0001) orientation but with a unit cell rotated around the [0001] axis by an angle of 30° in relation to the underlying sapphire to compensate for the different atomic arrangement [44–47]. The atomic arrangement of the rotated unit cells can be seen in Fig. 1.2a. The Al-Al distance of the AlN in [100] direction will adapt to the Al-Al distance of the sapphire in [1$\bar{1}$0] direction. This results in an effective in-plane lattice constant of the sapphire of $a_{\mathrm{Al_2O_3}}/\sqrt{3}$ to which the AlN with

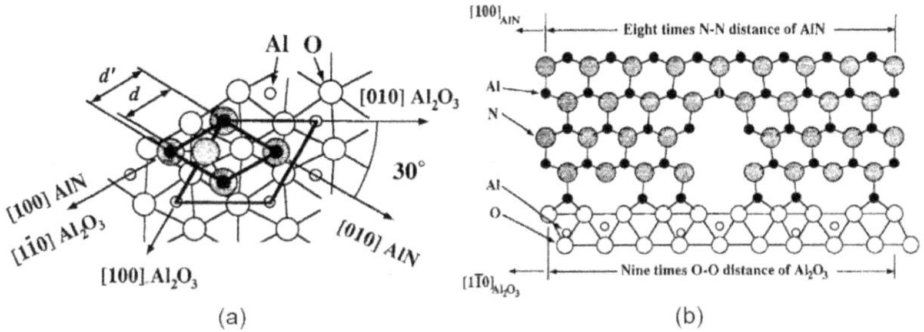

Figure 1.2: Illustration of the heteroepitaxial relation between AlN and sapphire in **(a)** plan view and **(b)** cross section. The AlN nucleates with a unit cell rotated by an angle of 30° with respect to the unit cell of the underlying sapphire. The resulting lattice mismatch of 13.29 % leads to 9 times the Al-Al distance of Al_2O_3 fitting to eight times the Al-Al distance of AlN combined with an edge dislocation in the center. From reference [48] with permission of AIP Publishing © 1994.

lattice constant a_{AlN} needs to adapt. This yields a lattice mismatch f of

$$f = \frac{a_{\mathrm{AlN}} - a_{\mathrm{Al_2O_3}}/\sqrt{3}}{a_{\mathrm{Al_2O_3}}/\sqrt{3}} = 13.29\,\%. \tag{1.1}$$

The lattice mismatch leads to growth according to nine times the Al-Al distance of Al_2O_3 fitting to eight times the Al-Al distance of AlN combined with an edge dislocation in the center depicted in Fig. 1.2b. This reduces the effective lattice mismatch after several monolayers.

Due to the large lattice mismatch AlN almost immediately grows in 3-dimensional growth mode on top of the sapphire leading to independent nucleation islands instead of a continuous 2-dimensional layer [49–51]. A 2-dimensional layer is formed only after subsequent AlN growth until coalescence of the nucleation grains is reached. The independent 3-dimensional nucleation grains lead to a certain distribution of tilt and twist between them. At the point of their coalescence, tilt and twist result in low angle grain boundaries which are compensated by the generation of a huge amount of threading dislocations (TDs) with a density above $1 \times 10^{10}\,\mathrm{cm}^{-2}$ [52–55]. Dislocations are one-dimensional line defects which are characterised by their Burgers vector $\vec{b}$ and line vector $\vec{l}$. The line vector l points along the core of the dislocation which is the center of the imperfection of the lattice. A TD names a dislocation for which the line vector has a component in growth direction so that the dislocation penetrates through the layer in growth direction. The Burgers vector can be determined by drawing a closed

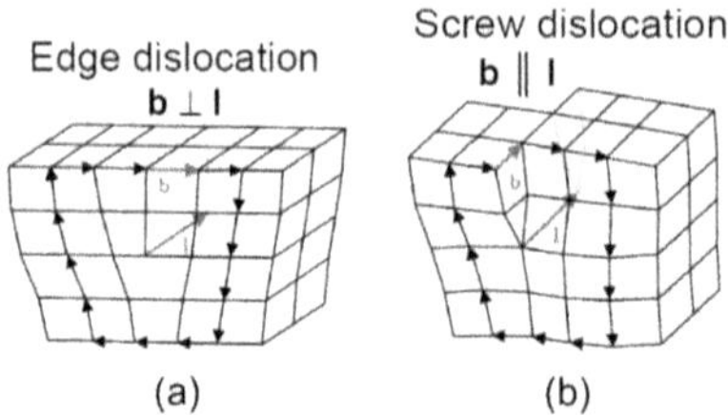

Figure 1.3: Sketch of **(a)** an edge dislocation and **(b)** a screw dislocation. An edge dislocation is associated with an extra half plane inside the crystal lattice, while a screw type dislocation is associated with a distortion of crystal planes to each other. From reference [56] with permission of Springer Nature © 2013.

path in the perfect lattice of a crystal and then again in the disturbed lattices with the closed path encircling the dislocation. The same path which was closed in the perfect lattice will be open in the lattice with the dislocation. The difference between the two closed paths is the Burgers vector. This method to determine the Burgers vector is visualised in Fig. 1.3 for the two basic kinds of dislocations, the edge dislocation and the screw dislocation. The edge dislocation is associated with an extra half plane inserted in the crystal lattice. The definition of an edge type dislocation is that the Burgers vector $\vec{b}$ is perpendicular to the line vector $\vec{l}$. The screw dislocation is associated with a distortion of the crystal planes to each other. The definition of a screw dislocation is that the Burgers vector $\vec{b}$ is parallel to the line vector $\vec{l}$. Geometrically, edge dislocations compensate twist of the nucleation grains, while screw dislocations compensate tilt. A dislocation for which the Burgers vector $\vec{b}$ is neither perpendicular nor parallel to the line vector $\vec{l}$ is called mixed dislocation.

If dislocations meet, they will react with each other under the condition that the overall Burgers vector is conserved [57]. This translates to the equation

$$\sum_i \vec{b}_i = \sum_j \vec{b}_j \,, \tag{1.2}$$

with $\vec{b}_i$ being the Burger vectors of the incoming dislocations and $\vec{b}_j$ being the Burgers vectors of the outgoing dislocations. For two dislocations with opposite Burgers vectors the sum over the Burgers vectors is zero. This process is called annihilation and results in an overall dislocation reduction.

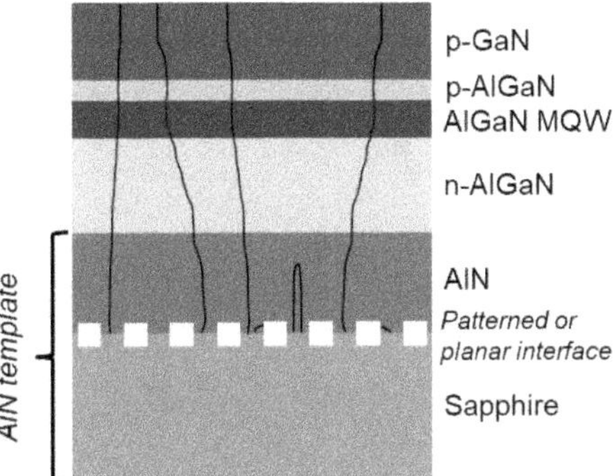

Figure 1.4: Simplified layer stack of a UV LED heterostructure including the AlN template with planar or patterned AlN/sapphire interface. The active region with the MQW lies between the n-side and p-side of the device. TDs (dislocations penetrating through the heterostructure) are indicated by the black lines. The light is usually extracted via the sapphire substrate due to the absorbing p-GaN.

1.2 AlN templates as UV LED base layers

The AlN templates consist of (0001) AlN grown on (0001) sapphire, also called *c*-plane AlN and *c*-plane sapphire. They are used as the foundation for subsequently deposited UV LED heterostructures. Usually, the AlN template is grown in a first epitaxial growth run separated from the second growth run to grow the UV LED heterostructure on top. In Fig. 1.4a the simplified layer stack of a typical UV LED heterostructure including the underlying AlN template is shown. In general, an LED heterostructure consists of the pn-junction and the quantum wells between the p- and n-side forming the active region of the LED [58–61]. In case of the UV LED heterostructure on top of the AlN template, the n-side is grown with silicon-doped AlGaN. The following active region is formed by the multi-quantum-well (MQW) stack whereas each quantum well is separated by the barriers. The p-side on top of the MQW stack starts with a magnesium-doped p-AlGaN capped by a magnesium-doped p-GaN layer. The light is generated by electrical contacting so that electrons and holes are injected via the n-side and p-side, respectively. Accumulation of electrons and holes in the MQW leads to increasing probability of radiative recombination in the MQW. This results in photon emission with the characteristic wavelength of the quantum wells. The crucial question is now, how the properties of the AlN template will influence the performance of the UV LED heterostructure. Before this can be evaluated the performance of an LED needs to be defined. The ratio between the input carrier pairs and the extracted photons is

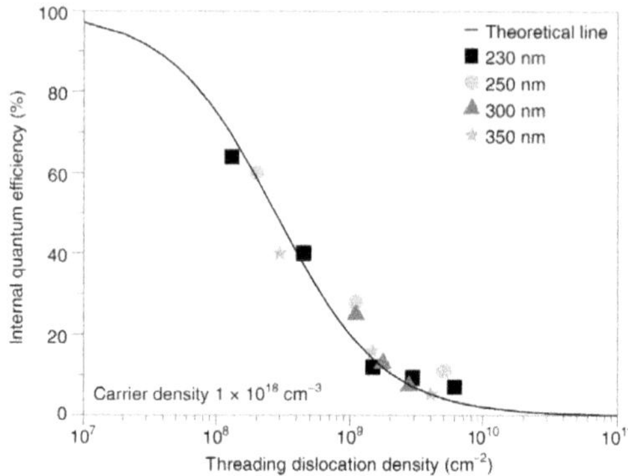

Figure 1.5: Simulated data combined with experimental results for the dependency between internal quantum efficiency and the TDD for a constant carrier density of $1 \times 10^{18}\,\mathrm{cm}^{-3}$. The experimental results are in good agreement with the cure predicted by the simulation. From reference [62] with permission of Springer Nature © 2019.

the external quantum efficiency η_{EQE} which fulfils the following relation

$$\eta_{\mathrm{EQE}} = \eta_{\mathrm{rad}} \cdot \eta_{\mathrm{inj}} \cdot \eta_{\mathrm{LEE}} = \eta_{\mathrm{IQE}} \cdot \eta_{\mathrm{LEE}}. \tag{1.3}$$

The radiative efficiency η_{rad} is the ratio between the radiative recombination inside the MQW and other for instance non-radiative recombination processes. The injection efficiency η_{inj} describes how efficient the injected electrons and holes reach the MQW and accumulate inside. The product of η_{rad} and η_{inj} is the internal quantum efficiency η_{IQE}. The ratio of the generated photons, which are extracted out of the LED heterostructure, in relation to all generated photons inside the MQW is the light extraction efficiency η_{LEE}. The AlN base layer has a great influence on all of these parameters. The focus of this work lies on the influence of the radiative efficiency η_{rad} and the light extraction efficiency η_{LEE}. The impact of the properties of the AlN templates on these parameters will be briefly discussed in the following.

Crystal quality As mentioned before, the crystal quality of the AlN base layers is mainly defined by the amount of TDs. TDs will penetrate through the whole layer stack as indicated in Fig. 1.4. Inside the MQW they will lead to non-radiative recombination which directly limits the radiative efficiency η_{rad}. In Fig. 1.5 the effect of the threading dislocation density (TDD) on the radiative efficiency is quantified assuming a constant carrier density of $1 \times 10^{18}\,\mathrm{cm}^{-3}$. Theoretical calculations of the internal quantum efficiency in dependency of the TDD in the active region are presented and compared with experimental results for UV LEDs of different wavelengths [63–69]. If the TDD is reduced from $1 \times 10^{10}\,\mathrm{cm}^{-2}$ down to $1 \times 10^{8}\,\mathrm{cm}^{-2}$ the IQE is predicted to increase from approx. 2 % to approx. 75 %. The reduction of the TDD can be considered as one of the most important tasks for achieving high performance LEDs. For pseudomorphic growth of the UV LED heterostructure on the AlN template at least up to the active region, the TDD in the active region is determined by the TDD of the underlying AlN template. The TDD will also have an influence on the injection efficiency η_{inj}. Usually, layers with lower TDD show better conductivity [70].

Surface morphology In terms of surface morphology it is important to provide a continuous and homogeneous surface, since usually any disturbance of the AlN surface will be directly translated to the subsequently grown AlGaN. Additionally, the layer should be reasonably smooth to suppress inhomogeneous AlGaN growth of different compositions. Inhomogeneous AlGaN growth results in broadening of the emission peak of the related UV LED. Furthermore, a rough AlN surface leads to rough AlGaN growth inducing smearing out of the MQW which decreases carrier localisation and thus, radiative recombination. Usually, the optimal condition for the AlN surface morphology is an atomically smooth surface, since this is a good basis for reproducible heterostructure growth. However, there also has been work of providing AlN base layers with a heavily step bunched surface. Inhomogeneous AlGaN incorporation leads to a broad peak with an increased integrated emission power [71]. However, in this work a narrow peakwidth of the final UV LEDs is desired.

AlN/sapphire interface Up to now, a p-doped GaN layer is normally used for electrical contacting at the p-side. This layer is absorbing in the UV range due to the bandgap of GaN being 3.4 eV which translates to a band edge of approx. 360 nm [72]. Hence, 50 % of the generated light will be lost in the p-side. The other 50 % of the generated light will propagate towards the transparent AlN template where part of it will be extracted. However, the refractive index profile of AlN and sapphire results in total reflection so that a huge portion of the light is reflected back into the structure and absorbed in the p-side. To soften this total reflection the shape of the AlN/sapphire interface plays an important role. By implementing a patterned AlN/interface, increased scattering has the potential of improving the light extraction efficiency. A patterned interface can be achieved by epitaxial lateral overgrowth. This also enables efficient defect reduction. However, the criterium of a smooth AlN surface is challenging to achieve, since the layer needs to properly coalesce.

Strain state The growth of the UV LED heterostructure on top of an AlN template is always accompanied by the formation of strain, since AlGaN has a larger in-plane lattice constant compared to AlN. If during growth the strain becomes too high, the layer will relax by for instance roughening or the formation of new TDs which becomes energetically favourable. For growth of UVA and UVB heterostructures, the Ga-content needed is too high for pseudomorphic. Thus, relaxation will be achieved in a controlled way by appropriate strain management to generate as few new TDs as possible. In contrast, UVC LED heterostructures consist of AlGaN with high Al-content so that the heterostructure usually can be grown pseudomorphically up to the active region without any relaxation. However, the process always works on the edge of relaxation, since only a small increase in Ga-content in the n-AlGaN layer leads to a considerable improvement of the electrical conductivity. Hence, it is obvious that the strain state of the AlN buffer always has to be considered in order to be able to appropriately handle the strain during subsequent AlGaN growth. For AlN base layers with different strain states resulting in slightly different lattice constants at growth temperature, the desired layer is the one with the biggest in-plane lattice constant, since AlGaN has a bigger lattice constant than AlN.

These four criteria will be discussed along the scientific results of this work. While in literature the main focus often only lies on the reduction of the TDD, in this work the other criteria will also be considered.

1.3 State of research

In this section the state of research regarding preparation of AlN base layers for UV LEDs will be outlined [73]. After giving a short overview of the general field of planar AlN growth, the focus lies on the specific research topics of HTA and growth on patterned substrates. These methods have gained increasing research interest in recent years and are also the main focus of this work. The relevant literature of these methods will be sorted in work that was published before the start of this work in June 2017 and after. What has to be noted is that it is often difficult to compare different research work, since methods for instance for the determination of the TDD strongly differ. While determination by for instance transmission electron microscopy has to be carefully examined regarding the thickness of the specimen and the incident angle of the electron beam, the measurement of the X-ray omega rocking curve full width at half maximum (XRC-FWHM) by X-ray diffraction (XRD) is a more reliable method to get a benchmark of the related TDD.

Planar AlN growth Planar AlN growth is here referred to as AlN growth on planar sapphire without any post-thgrowth annealing step, since this will be described later in more detail. AlN growth on sapphire has a long history with work dating back to before the year 2000 [74]. Besides MOVPE growth there has been work also using hydride vapour phase epitaxy [75, 76] and molecular beam epitaxy [77]. Using MOVPE the polarity control at growth start during nucleation is crucial for the quality of the subsequently grown AlN. The ratio of Al-polar to N-polar AlN is influenced by the amount of oxygen in the reactor atmosphere [78–81]. Also the preflow (trimethylaluminium or nitridation) prior to the growth plays an important role [82, 83]. After successful nucleation the main mechanism for TDD reduction is increasing the layer thickness to induce mutual annihilation of TDs. Unfortunately, the AlN on sapphire layer system usually exhibits a high amount of tensile strain during growth which leads to layer cracking at thin layer thickness if no strain management is applied [84]. The strain management is usually done by inducing alternating 2D to 3D growth transitions in the AlN layer resulting in alternating roughening and smoothing of the layer. This can be done by different techniques which by changing the growth conditions will create a multilayer system for instance by the ammonia pulse-flow method or by varying the V/III ratio etc. [66, 84–87]. AlN layer thickness of up to 4 µm and resulting TDDs in the range of mid $10^8\,\mathrm{cm}^{-2}$ can be reached by these methods.

Growth on patterned substrates The first reports about AlN growth on patterned sapphire substrates date back to 2006 [88, 89]. A trench pattern in the micrometre range was used and the TDD determined by plan-view TEM is presented to be below $10^7\,\mathrm{cm}^{-2}$ [89]. However, the related XRC-FWHM of 300″ for the 0002 reflection and 400″ for the 20-24 reflection suggest a TDD in the range of approx. $1 \times 10^9\,\mathrm{cm}^{-2}$ range. In the late 2000s and early 2010s the AlN growth on stripes was established by different groups with similar results [90, 91]. In recent years it was shown that in case of GaN it can be very beneficial to go for patterned sapphire substrates with patterns in the nanometre range (NPSS) [92, 93]. First demonstrations of overgrowing NPSS by AlN were presented by Dong et al. in 2013 [94, 95]. They used nanosphere lithography to prepare the hole type nanopattern on the sapphire surface and ended up with a TDD of

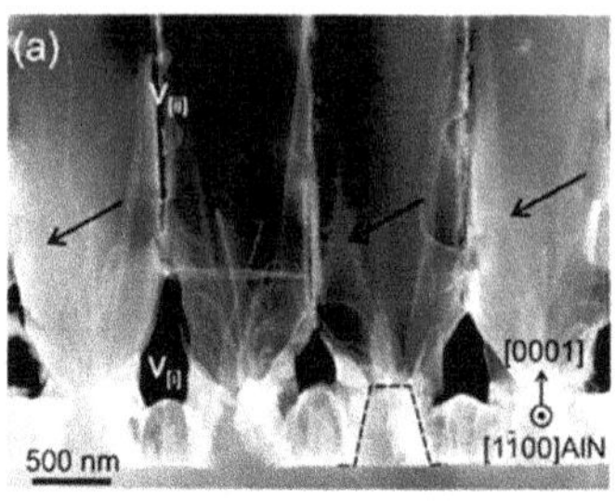

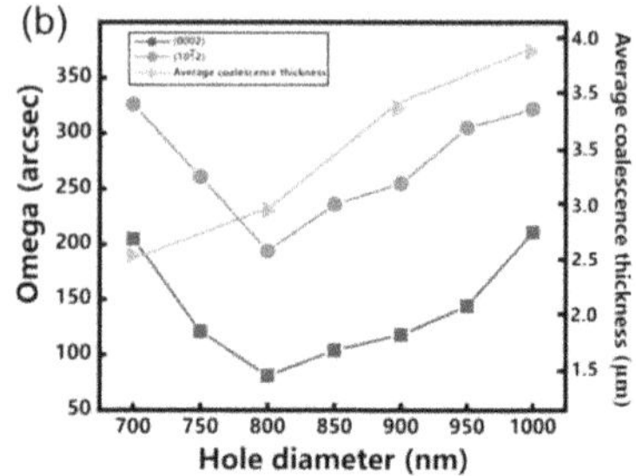

Figure 1.6: (a) TEM micrograph showing the growth start of AlN on nanopillar patterned sapphire with pattern pitch of 1 µm. From reference [97] with permission of John Wiley and Sons © 2016. **(b)** XRC-FWHM of the 0002 and 10-12 reflections and the average coalescence thickness against the hole diameter for AlN growth on nanohole patterned sapphire substrates with a constant pitch of 1.4 µm. From reference [100] with permission of The Japan Society of Applied Physics © 2019.

the subsequently grown AlN of $1.2 \times 10^9\,cm^{-2}$ estimated by calculation from the XRC-FWHM of 86″ for the 0002 reflection and 320″ for the 10-12 reflection. Subsequently grown UV LED heterostructures emitting at 282 nm showed an increase in the external quantum efficiency of 98 % compared to a flat sapphire substrate. More work from different groups followed with growth on nanoholes as well as nanopillars [96–99]. The lowest TDD achieved was as low as approx. $3.5 \times 10^8\,cm^{-2}$ determined by selective defect etching. This is in agreement with the XRC-FWHM of 171″ for the 0002 reflection and 205″ for the 10-12 reflection. The main reduction mechanisms are the bending of TDs to the free surfaces at growth start as well as the approximately defect free growth in the laterally overgrown regions as can be seen in the TEM image in Fig. 1.6a.

After June 2017 further work focused on further TDD reduction by finding the optimal pattern dimensions for reducing the area of coalescence in relation to the laterally overgrown regions, since coalescence fronts can lead to new TDs [100, 101]. In Fig. 1.6b the XRC-FWHM of the 0002 and 10-12 reflections and the average coalescence thickness are plotted against the hole diameter for a constant pattern pitch of 1.4 µm. For a hole diameter of 800 nm, XRC-FWHM of 162″ for the 0002 reflection and 181″ for the 10-12 reflection were measured which translates to a TDD of approx. $2.8 \times 10^8\,cm^{-2}$.

HTA of AlN/sapphire HTA of AlN/sapphire emerged in 2016. The group of Prof. Miyake at the Mie university in Japan published first results about this method in early 2016 [102–104]. They annealed a 300 nm thick AlN MOVPE layer on sapphire in a carbon-saturated N_2-CO atmosphere at 1700 °C for 1 h and observed a reduction

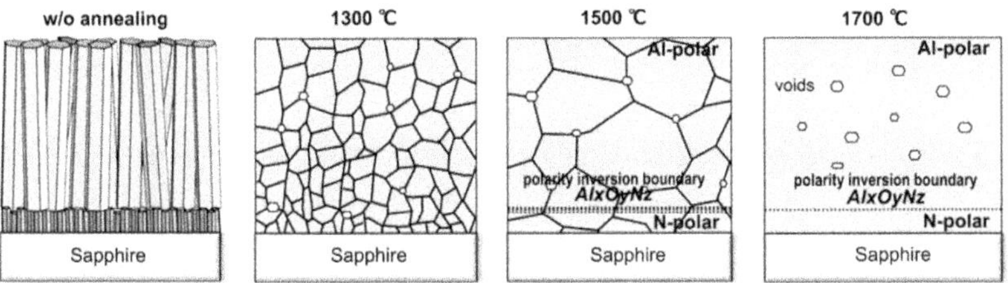

Figure 1.7: Sketch to explain the mechanism behind the TDD reduction by HTA. The group of Prof. Miyake suggest solid state epitaxy leading to gradual grain coalescence in the work of Xiao et al. as indicated in the sketch. From reference [110] with permission of Elsevier © 2018.

of the TDD by more than one order of magnitude down to $4.7 \times 10^8\,\mathrm{cm}^{-2}$ determined by plan-view transmission electron microscopy (TEM). Significant simplification of the method led to the publication of Miyake et al. in December 2016 which can be considered as the breakthrough in terms of impact in the UV LED community [105]. This was because the AlN layers for annealing were not prepared by MOVPE anymore, but by the much easier method of sputtering. Furthermore, nitrogen was now used as annealing environment. For stabilisation of the surface the layers were simply put face-to-face (epitaxially grown surface on epitaxially grown surface). XRC-FWHM as low as 49″ and 287″ were reached for the 0002 and 10-12 reflections, respectively. Successful AlN MOVPE regrowth was demonstrated in collaboration with Miyake and Huang et al. while conserving the material quality [106].

After June 2017 the successful use of such AlN templates as basis for UV LEDs was demonstrated by Susilo et al. in collaboration the group of Prof. Kneissl at the Technical University of Berlin, the group of Prof. Weyers at FBH, the company Evatec AG and the group of Prof. Miyake [107]. The research interest in HTA further increased resulting in a lot of research work being published in the following years mainly by the group of Prof. Miyake [108–125], but also by other research groups [126–130]. The mechanism for the TDD reduction is not yet fully understood. While Xiao et al. in the group of Prof. Miyake suggest solid state epitaxy at the elevated temperature resulting in gradual grain coalescence as indicated in Fig. 1.7 [119], other groups discuss vacancy-assisted dislocation climb resulting in dislocation annihilation [128,129]. Just recently a publication was released describing a double sputtering process based on HTA after each of the two sputter depositions [125]. The TDD was estimated to be $4.3 \times 10^7\,\mathrm{cm}^{-2}$ determined by plan-view TEM on an area of 8 µm x 8 µm. The corresponding XRC-FWHM were measured to be 12″ for the 0002 reflection and 70″ for the 10-12 reflection also suggesting a TDD in the mid $10^7\,\mathrm{cm}^{-2}$. In view of the results described above, this is the lowest TDD achieved and hence, an outstanding result.

The starting situation at FBH regarding the state of the art AlN templates is that two different technologies are established. 1.5 µm crack-free AlN can be prepared by growth on planar sapphire using the roughening and subsequent smoothing of the surface for strain management as described above. With this method TDDs of mid $10^9\,\mathrm{cm}^{-2}$ are achieved at a reasonable expense. For lower TDDs epitaxial lateral overgrowth of patterned AlN/sapphire with a stripe pattern of 3.5 µm pitch was developed. A crack-free AlN surface of 6 µm thickness and TDDs of $1\text{-}2 \times 10^9\,\mathrm{cm}^{-2}$ are obtained. However, the in-house patterning process of the AlN/sapphire surface combined with two epitaxial growth steps (before and after patterning) and a high layer thickness makes the technology very cost intensive. Hence, a simpler technology with similar or lower TDDs is desired.

CHAPTER 2

Experimental methods

Details about the experimental methods used in this work will be presented in the following chapter. For the basic knowledge the reader will be referred to standard literature. The scope of this chapter is to provide particular technical details for the methods applied in this work. Additionally, extended processing of data like the determination of the TDD from XRD measurements will be illustrated.

2.1 Sample fabrication

For sample fabrication, MOVPE is performed for AlN growth, sintering furnaces are used for HTA and the NPSS are processed by lithography and subsequent dry etching. Details will be given in the following.

MOVPE reactor

Metalorganic vapour phase epitaxy is one of the main techniques for growth of III-V semiconductor thin films. The physical principles of MOVPE growth and the specific MOVPE growth of nitride semiconductors are described in references [56,131,132]. The MOVPE reactor for preparation of the AlN base layers is depicted in Fig. 2.1. It is an AIX2400G3HT™planetary reactor from Aixtron with a capability of 11 x 2 inch wafers and a high temperature setup. For the high temperature stability of the reactor parts up to process temperatures of approx. 1400 °C, the susceptor and satellites made of graphite are coated with tantalum carbide (TaC). The process temperature is measured at the backside of the heated susceptor by pyrometry. The surface of the wafers, which

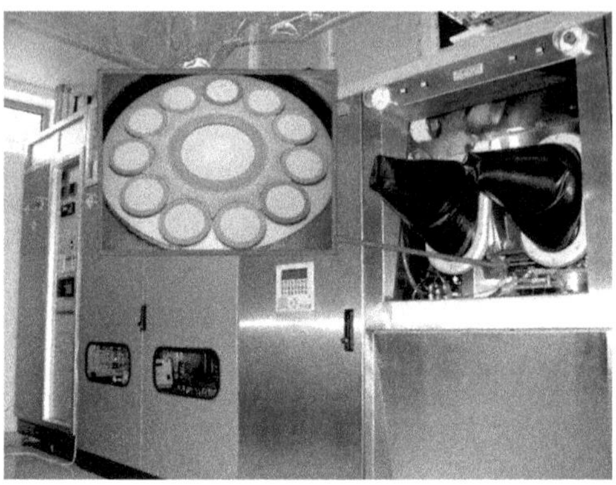

Figure 2.1: Planetary MOVPE reactor AIX2400G3HT™with capability of 11 x 2-inch wafers. The inset shows the susceptor with the 11 satellites working as substrate holders inside the reactor. The precursor gases flow horizontally over the wafers on the satellites.

will be heated by heatcoupling to the satellites and susceptor, will be approx. 50 °C to 100 °C lower than the process temperature. For AlN growth the precursor gases trimethylaluminium (TMAl) and ammonia (NH_3) are used with hydrogen (H_2) as the carrier gas. The gases are introduced from a central nozzle above the susceptor and flow radially to the exhaust at the reactor sidewalls resulting in a horizontal flow over the wafers. Rotational symmetry is improved by rotation of the susceptor and each satellite in opposite directions. All growth experiments were performed at a total pressure of 50 mbar on 2-inch sapphire substrates with offcut of 0.2° or 0.1° towards the m direction. Growth experiments were performed by varying the process temperature and the V/III ratio. The V/III ratio in case of AlN growth is the input ratio of the fluxes of NH_3 and TMAl. However, what has to be kept in mind is that the effective V/III ratio at the wafer surface during growth remains unknown, since e.g. prereactions in the gas phase and the flow pattern will alter the effective V/III ratio. The effective V/III ratio depends heavily on the reactor design. Hence, it is important to note that the developed growth processes are not simply applicable to other MOVPE reactors. The particular growth parameters used for the different growth experiments will be given at the relevant passages in the following chapters.

The MOVPE growth of the UVC LED heterostructures with a targeted wavelength of 265 nm presented in chapter 5 was performed at the Technical University of Berlin in a close coupled showerhead reactor [133–135]. The usual precursors trimethylgallium (TMGa), triethylgallium (TEGa), TMAl, NH_3, biscyclopentadienylmagnesium (Cp_2Mg)

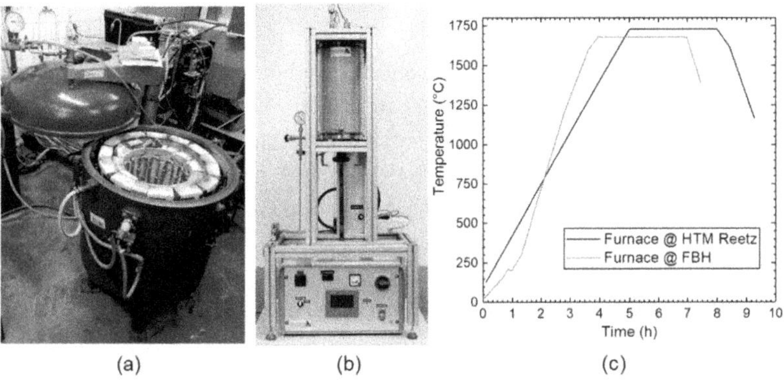

Figure 2.2: Setup of the **(a)** annealing furnace at HTM Reetz and **(b)** annealing furnace at FBH. **(c)** Temperature transients for exemplary annealing experiments at 1730 °C for 3 h for the HTM Reetz furnace and 1680 °C for 3 h for the FBH furnace.

and silane (SiH_4) with hydrogen and nitrogen as carrier gas were used. For growth of the standard UVC LED heterostructure with fully absorbing p-side, another 400 nm AlN layer was grown homoepitaxially on top of the provided AlN base layers. The 900 nm $Al_{0.76}Ga_{0.24}N$:Si current spreading layer and 200 nm $Al_{0.65}Ga_{0.35}N$:Si n-contact layer was then grown with a grading in the Al-content between the layers of 100 nm thickness followed by the active region consisting of an $Al_{0.48}Ga_{0.52}/Al_{0.63}Ga_{0.37}$ MQW stack. For the electron blocking layer a 10 nm thick $Al_{0.85}Ga_{0.15}N$ interlayer and 25 nm thick $Al_{0.75}Ga_{0.25}N$:Mg was grown, while the final capping layer consists of 200 nm GaN:Mg working as the p-contact layer. For growth of the UVC LED heterostructures with transparent p-side, the heterostructure has an identical design up to the electron blocking layer. However, to achieve transparency of the p-side for the targeted wavelength of 265 nm instead of the GaN contact layer, a short period super lattice of $Al_{0.65}Ga_{0.35}N/Al_{0.71}Ga_{0.29}N$ with average composition of $Al_{0.68}Ga_{0.32}N$ is grown on top.

Annealing furnaces

Two different hot wall furnaces were used for HTA of AlN/sapphire layers in face-to-face configuration. Both are shown in Fig. 2.2. The annealing furnace in Fig. 2.2a is located at the company HTM Reetz in Berlin. It was used until the annealing furnace in Fig. 2.2b was put into operation at FBH. HTA was applied at a pressure of 1000 mbar in nitrogen ambient for both furnaces with the temperature being controlled by a thermocouple at the respective heater. However, the temperature ramps for heating up and cooling down are different. In Fig. 2.2c temperature transients for specific annealing experiments are exemplarily plotted for both furnaces. For the FBH furnace the temperature curve is shown for HTA at 1680 °C for 3 h. For the HTM Reetz furnace the temperature curve is shown for HTA at 1730 °C for 3 h. The FBH furnace is heated with 10 K min^{-1} up to 1400 °C followed by 8 K min^{-1} up to 1600 °C and 4 K min^{-1} up to 1680 °C. The cooling down after 3 h of holding time at 1680 °C happens with 10 K min^{-1} down to 1400 °C. The HTM furnace is heated with a constant rate of approx. 5.3 K min^{-1} up to 1730 °C. After 3 h of holding time at 1730 °C, the cooling down to 1400 °C happens with 7.4 K min^{-1}. Both, the heating up and the cooling down happens faster for the FBH furnace. In case of equal annealing parameters of 1680 °C for 3 h, this would result in an increased time of 16 min above a temperature of 1400 °C in case of the HTM Reetz furnace (268 min instead of 252 min). The temperature of 1400 °C can be considered as the particular point above which annealing effects become significant, since the temperature during MOVPE growth starts to be significantly exceeded. For measuring the actual temperature close to the samples, another thermocouple was placed inside each furnace. This thermocouple additional to the one at the heater measured a temperature of 1715 °C for the FBH furnace with the temperature control being at 1680 °C and 1780 °C for the HTM Reetz furnace with the temperature control being at 1730 °C. So for both furnaces the actual temperature of the samples is higher than at the heater. The temperature difference is 35 °C for the FBH furnace and 50 °C for the HTM furnace. Hence, for nominally equal annealing experiments the HTM Reetz furnace will be more hot additional to the longer annealing time above 1400 °C.

Processing of nanopatterned sapphire substrates

The NPSS used in this work were fabricated by different methods. For an exemplary description of a specific fabrication procedure, the method developed at the University of Bath is presented. With the University of Bath an extended collaboration took place during this work with the possibility to participate in the fabrication of one batch of wafers to gain deeper insights into the method. The rather new and interesting technique of Displacement Talbot Lithography (DTL) was used for lithography. The advantage of this method lies in the potential of large scale and low cost lithography of patterned structures with small dimensions [136, 137]. Using DTL and subsequent dry etching, NPSS consisting of nanopillars and nanoholes were processed. More details of the fabrication process will be given in the following. The process flow for fabrication

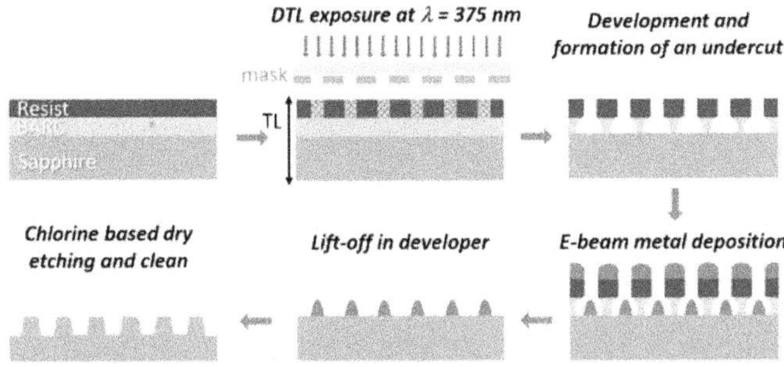

Figure 2.3: Process flow for fabrication of sapphire nanopillars. After exposure of the BARC and resist and subsequent development, nickel is deposited to form an array of nanodots on the bare sapphire. The rest of the BARC and resist is removed and the pattern of the nanodots is transferred into the sapphire by ICP etching using the nickel dots as mask material. To complete the process the residual nickel dots are removed by aqua-regia. From reference [134] (first authorship) with permission of Elsevier © 2020.

of nanopillar NPSS with pitch of 1 µm is shown in Fig. 2.3. Prior to DTL, a bottom antireflective coating (BARC) layer (WiDE 30 W – Brewer Science) and a high-contrast positive resist layer (Dow Ultra-i 123 diluted with Dow EC11 solvent) was spin-coated on 2-inch sapphire substrates. The resist was exposed by DTL (PhableR 100, Eulitha) through an amplitude mask with a hexagonal configuration of 550 nm diameter circular openings with a pitch of 1 µm. After DTL and development, nanoholes were generated in the resist with an undercut in the BARC layer beneath the resist. The following metal (nickel) deposition by e-beam evaporation and subsequent lift-off by further developing produces an array of nickel nanodots on the bare sapphire. The nickel nanodots were used as the mask for etching so that by inductively coupled plasma (ICP) dry etching (Oxford Instruments System 100 Cobra) the pattern of the nanodots could be transferred into the sapphire. $Cl_2/BCl_3/Ar$ chemistry was applied for etching. The residual nickel nanodots were removed by soaking in aqua-regia solution ($HCl{:}HNO_3$, 3:1). The nanopillars with pitch of 1 µm were checked to be uniform across the whole 2-inch wafer.

For fabrication of the nanoholes with pitch of 1 µm the process flow follows the sketch shown in Fig. 2.4. In contrast to fabrication of the nanopillars, a SiN layer is deposited on the sapphire before the BARC and resist layer. The SiN will be used as the hardmask for the dry etching. By DTL exposition and subsequent development of the resist, circular openings are generated in the resist layer. These openings are transferred into the SiN hardmask by ICP etching with CHF_3 chemistry. The rest of

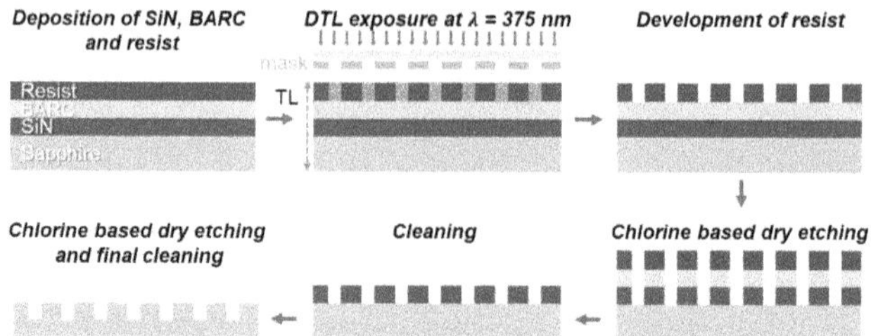

Figure 2.4: Process flow for fabrication of sapphire nanoholes. The BARC and resist on top of the SiN are exposed to DTL. After development the nanohole pattern is generated in the resist and transferred to the SiN hardmask by ICP etching. The pattern is then transferred into the sapphire by another ICP etching step before the surface is cleaned to remove the residual SiN.

the BARC and resist is removed by further developing before the pattern is transferred from the SiN hardmask into the sapphire by ICP etching with $Cl_2/BCl_3/Ar$ chemistry. To complete the process the surface is cleaned with buffered oxide etch (NH_4:HF = 5:1). The resulting nanohole pattern with 1 µm pitch shows uniformity across the whole 2-inch wafer.

2.2 Characterisation methods

The characterisation methods are divided into two categories. The basic methods are usually applied for every growth experiment to get a first picture of the outcome in terms of for instance surface smoothness, material quality etc. If more specific questions arise, advanced methods are employed mostly in collaboration with other researchers who are experts in these methods.

Basic methods

In-situ 405 nm reflectance To measure the 405 nm reflectance of the sample in-situ during MOVPE growth enables the evaluation of the growth rate and whether the layer is growing smooth or rough. A LayTec EpicurveTT system was used for this. An average 405 nm reflectance of approx. 0.13 indicates a smooth layer of AlN on sapphire. Hence, for instance the point of coalescence of an AlN surface can be estimated for growth on patterned substrates or intentional roughening of the layer for strain relaxation can be controlled efficiently. Additionally, Fabry-Pérot oscillations due to alternating constructive and destructive interference inside the layer stack emerge,

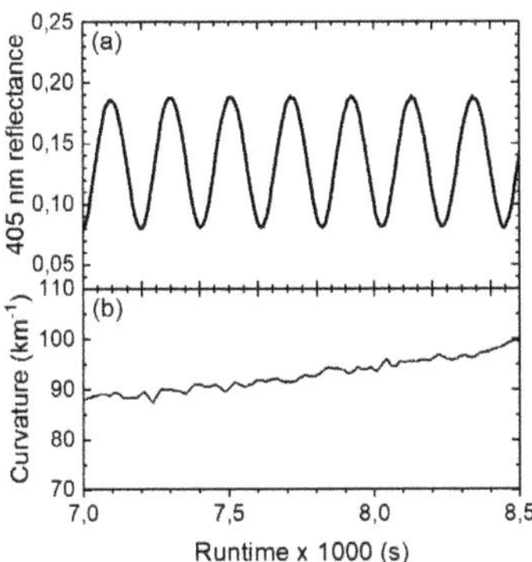

Figure 2.5: **(a)** 405 nm reflectance and **(b)** curvature against runtime for growth of AlN.

which are associated with a certain layer thickness. In that way the growth rate can be determined by a fit calculated in the LayTec EpiNet software, which takes the temperature dependent optical constants of the layer stack into account. One full Fabry-Pérot oscillation for AlN on sapphire at typical growth rates and a wavelength of 405 nm translates to approx. 100 nm layer thickness. Exemplarily, the 405 nm reflectance against runtime for growth of AlN at growth temperature of 1180 °C is shown in Fig. 2.5 for which the Fabry-Pérot oscillations can be seen.

In-situ curvature The in-situ curvature can be used as an indirect measure for the strain of the layer during growth. It is measured with a LayTec EpicurveTT system via three parallel laser beams, which are focused on the wafer surface with a certain initial distance. By measuring the positions of the reflected beams, the change of the distance for initial and reflected beams can be converted to the curvature of the wafer. Details of the experimental setup are found in reference [138]. The curvature κ of the wafer is related to the stress σ of the thin layer on the thick substrate via Stoney's equation [139]:

$$\sigma = \frac{E_S h_S^2 \kappa}{6 h_l (1 - \nu_S)}, \tag{2.1}$$

with E_S being the Young's modulus of the substrate, h_S the thickness of the substrate, h_l the thickness of the layer and ν_S the Poisson's ratio of the substrate. The stress of the layer is proportional to the strain of the layer with the proportionality factor being the Young's modulus of the layer E_l. In principal it is even possible to quantitatively

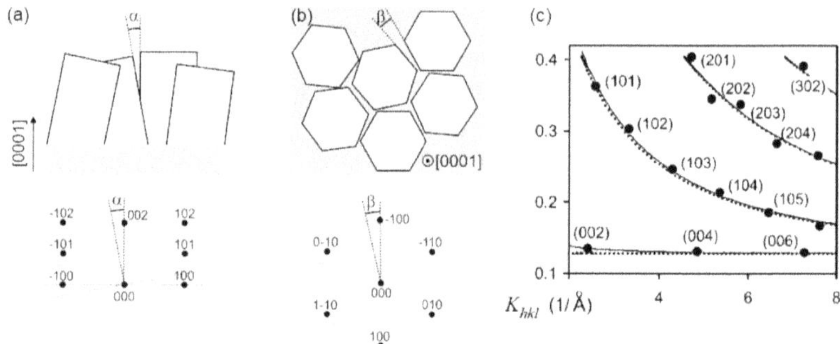

Figure 2.6: (a) Sketch for illustration how tilt and twist of grains in the layer lead to broadening of the reciprocal lattice points. The directions of broadening are different for tilt and twist. From reference [141] with permission of Cuvillier 2015. **(b)** XRC-FWHM of many different reflections measured for AlN on sapphire. The distribution of the XRC-FWHM is well described by the fit which is related to the eq. 2.6 with omission of the last term. From reference [142] with permission of AIP Publishing © 2005.

analyse the strain of the layer by processing of the curvature data [140]. However, in this work the curvature is only used as a qualitative measure for the strain of the AlN layer during growth on sapphire. Exemplarily, the curvature against runtime for growth of AlN at growth temperature of 1180 °C is shown in Fig. 2.5. The curvature is increasing which indicates the build-up of tensile strain.

X-ray diffraction XRD is the most powerful basic method used for the characterisation of the AlN base layers. The basics of XRD on semiconductors and the application to III-nitrides can be found in the references [143, 144]. In this work the main application of XRD is the determination of the crystal quality of the AlN base layers. This can be done by measuring the XRC-FWHM. The XRC-FWHM is directly linked to the perfection of the alignment of the crystal planes. For a perfect crystal, constructive interference would only be present under singular angles at which the Laue condition

$$\vec{k}_{\text{in}} - \vec{k}_{\text{out}} = \vec{G} \tag{2.2}$$

is fulfilled, with $\vec{k}_{\text{in}}$ being the wavevector of the incident beam, $\vec{k}_{\text{out}}$ the wavevector of the diffracted beam and $\vec{G}$ a reciprocal lattice vector. In that case the reciprocal lattice points made visible by XRD would be point-like. Every imperfection of the crystal lattice leads to a broadening of the reciprocal lattice points, since constructive interference is fulfilled for angles slightly off, which are related to disordered crystal planes. This is indicated in the sketch in Fig. 2.6. Tilt/twist angles of grains inside a single crystal film lead to a broadening of the reciprocal lattice points of the overall layer. The main imperfections in the investigated AlN base layers are TDs. Hence, larger FWHM are related to a higher amount of TDs. As described in chapter 1.1, screw dislocations are associated with tilt, while edge dislocations are associated with twist of the grains. It is possible to give an estimation of the amount of TDs by calculation from the XRC-FWHM. The basic principle following the references [142, 145–147] is outlined in the following to derive the formula utilised in this work.

As mentioned before, the broadening of the reciprocal lattice point is directly linked to the distribution of the orientation of the grains in the layer. Starting from this assumption, Gay et al. derived a relation between the FWHM β of this distribution to the TDD ρ under the assumption of a Gaussian distribution regarding the orientation of the grains:

$$\rho = \frac{\beta}{bt\sqrt{2\pi \ln 2}}, \tag{2.3}$$

with b the burgers vector and t the spacing between the grain boundaries. Dunn et al. simplified this formula under the assumption of randomly distributed dislocations to

$$\rho = \frac{\beta^2}{4.35b^2}. \tag{2.4}$$

The next step is to isolate the tilt component β_{tilt} and twist component β_{twist} of the XRC-FWHM β. Tilt is related to screw dislocations, while twist was related to edge dislocations. Since edge and screw type dislocations show different Burgers vectors in the wurtzite lattice of AlN, their densities need to be calculated independently. The broadening of the symmetric reflections is attributed only to tilt, since symmetrical reflections show rotational symmetry. Hence, by taking the XRC-FWHM of the 0002 reflection for β_{tilt} and the Burgers vector for the screw type dislocations b_c=0.4982 nm [146], the following formula for calculation of the screw dislocation density ρ_{screw} is found:

$$\rho_{\text{screw}} = \frac{\beta_{0002}^2}{4.35b_c^2}. \tag{2.5}$$

For calculation of the edge type dislocation density, the twist component β_{twist} of the XRC-FWHM β needs to be isolated. Unfortunately, this is not possible analogue to the former case, since the X-ray beam would need to be diffracted at the side of the layer which is of only several micrometre thickness. Therefore, Lee et al. derived a general formula how to split up β in β_{tilt} and β_{twist}. The formula reads

$$\beta_{hkl} = (\beta_{\text{tilt}} \cos \chi)^2 + (\beta_{\text{twist}} \sin \chi)^2 + \frac{(\frac{2\pi}{L})^2}{K_{hkl}^2} \tag{2.6}$$

for Gaussian shape of the peaks and with hkl the Miller indices for the specific reflection, χ the angle between the surface normal of the crystal and the diffraction planes, L the lateral coherence length and K_{hkl} the reciprocal lattice vector. Fortunately, the latter term, which is not straightforward to access, can be omitted with negligible errors. This was checked by Lee et al. by measuring many reflections and fitting eq. 2.6 to the experimental results which is shown in Fig. 2.6b. The observed error was always below 20 %. In this work the same assumption is made so that a final formula can be derived for the edge type dislocations similar to the screw type dislocations using the 10-12 reflection, taking the 0002 FWHM as β_{tilt} and the Burgers vector for the edge type dislocations b_a=0.3112 nm [146]:

$$\rho_{\text{edge}} = \frac{\sqrt{\frac{\beta_{10\text{-}12} - (\beta_{0002}^2 \cdot \cos \chi)^2)}{\sin \chi^2}}}{4.35 b_{\text{a}}^2}. \tag{2.7}$$

The total TDD is then the sum of ρ_{screw} and ρ_{edge}. However, several things have to be kept in mind using this formula. Firstly, the mixed type dislocations were not considered in the calculation above. They will be counted twice, since they contribute to both β_{tilt} and β_{twist}. However, this is not expected to significantly affect the result, since the dominant part of dislocations in the AlN base layers will be edge type dislocations [148]. So the total TDD is mainly determined by the amount of pure edge type dislocations. Furthermore, the broadening of the XRC-FWHM is only ascribed to dislocations. This will never be true, since other broadening due to the instrument, wafer curvature etc. will always be present [144]. This can lead to an overestimation of the TDD. Finally, it has to be considered that along the way of deriving the formula for determining the TDD from the XRC-FWHM many assumptions were made. These might only be valid for specific conditions. So the result always has to be taken carefully as a rough estimation. However, from the experience made in this work, the TDDs calculated by this method mostly agree well TDDs determined by other methods as long as no other obvious broadening mechanisms are present.

The strain of the layer at room temperature can also be assessed by XRD via determination of the lattice constants. It is important to distinguish between the strain at growth temperature and at room temperature, since they will strongly differ due to the thermal mismatch of AlN and sapphire. For determining the lattice constants high resolution XRD is used for better accuracy. With this method the absolute angular positions of several symmetrical reflections are measured. From the angular positions of the symmetrical reflections the c lattice constant can be calculated according to

$$c = \frac{\lambda l}{2 \sin \theta}, \tag{2.8}$$

with $\lambda = 0.154\,056\,\mathrm{nm}$ the wavelength of the X-ray source (Cu $K_{\alpha 1}$), the Miller index l for the symmetric reflection and θ the angle of the incident beam. The strain in z-direction ϵ_{zz} can then be calculated as

$$\epsilon_{zz} = \frac{c_0 - c}{c}, \tag{2.9}$$

with c_0 being the natural lattice constant of AlN. Finally, under the assumption of a biaxially strained lattice the in-plane strain is determined using the relation

$$\epsilon_{xx} = \frac{\frac{E}{2\nu} + C_{13}}{C_{11} + C_{12}} \epsilon_{zz}, \tag{2.10}$$

with Young's modulus $E = 308\,\mathrm{GPa}$, Poisson's ratio $\nu = 0.179$ [149] and the elasticity constants $C_{11} = 396\,\mathrm{GPa}$, $C_{12} = 137\,\mathrm{GPa}$, $C_{13} = 108\,\mathrm{GPa}$ [150].

Atomic force microscopy Atomic force microscopy (AFM) is a well established technique for determining the topography of semiconductor surfaces. Details regarding the setup and measurement principle can be found in reference [151]. In this work it is used for studying the microscopic surface morphology of AlN on sapphire in areas up to 10 µm x 10 µm. The vertical resolution of the used instruments are in the sub nm range. Hence, the bilayer steps of atomically smooth AlN can be detected. Since the microscopic surface morphology is a crucial category for the quality of an AlN base layer, AFM measurements are performed as a standard measurement.

Light microscopy Light microscopy imaging is used to evaluate the macroscopic surface morphology. For macroscopic surface disturbances, e.g. layer cracking (compare chapter 3.1 Fig. 3.3), AFM is not suitable, since detection is very unlikely for usual AFM scan areas up to 50 µm x 50 µm. Only by combining AFM and light microscopy imaging, the surface morphology of the entire wafer can be properly examined.

Advanced methods

Scanning electron microscopy Scanning electron microscopy (SEM) is used for high resolution images of the semiconductor surface with resolution in the sub µm range. Details of the technique are found in reference [152]. For this work only AlN surfaces and sapphire surfaces were investigated. Besides the simple plan view imaging, some samples were cut into pieces and images of the cross section were acquired to study the structural evolution during growth. Inside an SEM also other characterisation methods can be conducted, for instance cathodoluminescence and electron channelling contrast imaging [153–155]. These methods were also used in this work and are described in the following.

Cathodoluminescence Cathodoluminescence (CL) works in the way that semiconductor material is excited by the energy of the electron beam inside an SEM. This results in emission of photons with characteristic wavelengths corresponding to energy transitions of the semiconductor. These photons are measured by an additional detector which determines wavelength and intensity of the emitted photons with high sensitivity. Due to the scanning of the electron beam across the surface area high spatial resolution is achieved. In the setup used in this work the samples can be cooled down to 80 K. This improves the radiative recombination inside the semiconductor material since nonradiative recombination induced by thermal phonons is suppressed. CL was used in this work to study the surface of AlGaN MQWs grown on AlN base layers to check their homogeneity. Additionally, darkspot densities can be determined, which are directly related to the TDDs, because TDs act as non-radiative recombination centers. Hence, the darkspots appear dark for all wavelengths.

Electron channelling contrast imaging With electron channelling contrast imaging (ECCI) it is possible to visualise dislocations by use of an SEM. Analogue to XRD also the electrons of the electron beam will be diffracted at the crystal planes. By tilting the sample, the diffraction conditions can be altered, since the angle between the sample surface, electron beam and detector of the backscattered electrons will be changed. With this procedure diffraction conditions can be found for which the dislocations appear as point-like contrast, since diffraction at these points will be different due to the disordered crystal planes [156,157]. One of the advantages of this method is that it is non-destructive and more precise than the estimation from XRD. Furthermore, it offers spatial resolution of single dislocation.

Transmission electron microscopy Transmission electron microscopy is a very powerful technique for studying semiconductor thin films. Basic principles are found in reference [158]. In this work several TEM techniques were used to study the AlN base layers. With high-angle annular dark field scanning TEM (HAADF-STEM) dislocations and many kinds of contrast-generating lattice properties, e.g. inversion domain boundaries, can be made visible. Under certain diffraction conditions it is possible to induce contrast only for specific types of dislocations. For diffraction conditions along the 0002 zone axis, the screw and mixed type dislocations are visible, while for diffraction conditions along the 11-20 zone axis, only edge and mixed type dislocations are visible. The annular dark field (ADF) measurement condition works similar making the contrast of all dislocations visible. With high resolution TEM (HRTEM) single atom columns can be resolved. This enables the identification of the arrangement of the atoms by comparing to simulated contrast arrangements using specific software.

Energy dispersive X-ray spectroscopy With energy dispersive X-ray spectroscopy atoms can be excited with very high energies so that they will emit X-rays with their characteristic wavelength different for each element. In that way the elemental composition of material at specific points can be estimated. In this work the method was used inside the TEM for determining the AlGaN composition on particular positions of a specimen studied in the TEM.

Defect-selective etching Defect-selective etching (DSE) is used for the determination of the TDD of the AlN base layers. The surface is etched by an alkaline melt resulting in etch pits at the positions of the TDs [159, 160]. After etching, SEM images are acquired to count the number of etch pits which directly corresponds to the number of TDs. The challenge of this method is to find appropriate etching times and temperatures to, on one hand, prevent the surface from being overetched so that the etch pits overlap and can not be counted anymore and, on the other hand, being sure that the surface is sufficiently etched so that at every TD there is an accompanied detectable etch pit. However, in contrast to ECCI it is a destructive method.

Secondary ion mass spectrometry With secondary ion mass spectrometry (SIMS) impurity concentrations inside semiconductor layers can be determined. The method works with an ion beam sputtering through the layer and a mass spectrometer to identify the elemental composition of the removed material. Details of the method can be found in reference [161]. Very low concentrations down to approx. $1 \times 10^{16}\,\mathrm{cm}^{-3}$

can be detected.

Transmission For measuring the transmission through in this case AlN base layers on sapphire substrates a setup covering a spectral range from 200 nm to 1100 nm by emission from a deuterium and halogen lamp is used. External light sources can be neglected by acquiring a reference spectrum prior to the actual measurement. Unfortunately, the setup of the transmission measurement system does not allow to collect light which is scattered inside the layer system for instance at the usually rough sapphire backside. This excludes a quantitative analysis of the transmission data. However, qualitatively absorption bands and the band edge can still be recognised.

Electroluminescence Electroluminescence (EL) is the measurement technique to estimate the performance of the UVC LED heterostructures grown on top of the AlN base layers. Current is injected via indium dots working as contacts. For the p-contact an indium dot is simply pressed on top of the surface of the LED heterostructure. For the n-contact the surface is carved down to the n-side by a needle at the wafer edge and the indium dot placed there. In that way current can be injected which will lead to light emission from the LED heterostructure. The light is collected in a spectrometer as a function of the wavelength and integrated over the main emission peak to obtain the emission power.

2.3 Method for light extraction simulations

The wave optical simulations presented in chapter 4.1 are used to calculate the potential benefit of the performance of UVC LEDs in terms of light extraction efficiency by implementing a nanopatterned AlN/sapphire interface. The simulation setup is shown in chapter 4.1 in Fig. 4.1 and the basic physical boundary conditions are described there. In this part some technical details of the simulation method will be given. Further details are found in reference [162].

In the most basic way, the simulation method consists of calculating the scattering matrices which describe the light propagation at material interfaces [163]. For the UVC LED layer stack, the whole AlGaN heterostructure is implemented as an AlN layer. Furthermore, only the transmittance from the AlN into the sapphire is studied. Thus, the AlN/sapphire interface is the only relevant interface. The point light source is put far away from the interface so that there is no direct excitation of modes that propagate purely in the x-y plane of the interface (evanescent modes). While the interface could

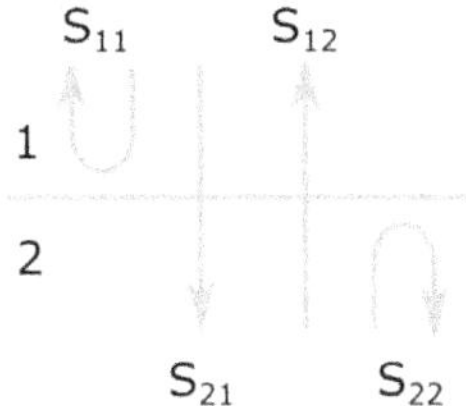

Figure 2.7: The relevant four light propagation cases for the AlN layer 1 and sapphire layer 2. S_{11} describes reflection from above, S_{21} tranmission through the interface from above, S_{22} reflection from below and S_{12} transmittance from below. From reference [162] (second authorship) with permission of The Optical Society (USA) © 2020.

scatter light into such modes, since both materials are assumed non-absorbing, these modes will decay radiatively into modes propagating in the z direction. Due to this, it is reasonable to neglect the evanescent modes in this case. This is in contrast to plasmonic nanostructures, where evanescent modes are a significant loss mechanism. In case of neglecting the evanescent modes, only light propagation with a component in the z-direction perpendicular to the interface remains. For each case of light propagation (reflection at the interface from above, transmittance through the interface from above, reflection at the interface from below and transmittance through the interface from below), a related scattering matrix exists as is schematically shown in Fig. 2.7. The respective entries in the scattering matrices are called coupling coefficients and need to be determined. The finite element method (FEM) is used for this with periodic boundary conditions in the x-y plane due to the periodicity of the nanopatterned interface. As input for the light propagation, plane waves are taken with two linearly independent light polarisations and multiple incident angles. The wavelength was set to 265 nm and the refractive indices used for AlN and sapphire are 2.35 and 1.83, respectively [164,165]. In the whole layer stack no optical losses were implemented. From convergence tests an accuracy of at least 10^{-2} was derived. The mesh length was set to 0.25 fold the wavelength of the light in AlN and sapphire. The finite element degree was designed adaptively in order to achieve an accuracy of 10^{-2} for every element [166]. Usually, a finite element degree of 3-4 was necessary. By carefully choosing the modes of the propagating plane waves, a lot of computational time can be saved. Firstly, if evanescent modes are excited, they will decay radiatively. Therefore, instead of modeling coupling of incoming radiation modes to evanescent modes and subsequent coupling of evanescent modes to outgoing radiation modes, only coupling from incoming radiation modes to outgoing radiation modes is considered. In the reciprocal k-space representation, this

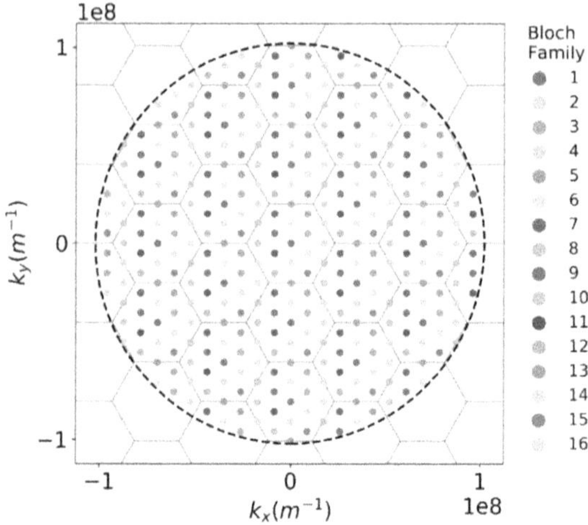

Figure 2.8: Representation of the scattering problem in reciprocal k-space. The relevant wave vectors lie inside the circle, which consists of many Brillouin zones of the nanopatterned interface. The dots with equal colours inside the Brillouin zones refer to wave vectors of the same Bloch family. From reference [162] (second authorship) with permission of The Optical Society (USA) © 2020.

translates into a cutoff wave vector $\vec{k}_c$ so that a circle in reciprocal space is formed with radius $\vec{k}_c$. Secondly, due to the periodicity of the nanopatterned interface the different $\vec{k}$ values can be further reduced. In reciprocal space the nanopattern translates into a periodic lattice of Brillouin zones related to the nanopattern. The crucial aspect about the Brillouin zone of a lattice is that equally located points in different Brillouin zones are connected by a simple relationship: One is obtained from the other by multiplying by a Bloch wave. Hence, we can reuse parts of the calculation for light incident with in-plane $\vec{k}$ vectors separated by a Bloch vector. In terms of calculation this means that the system matrix needs to be inverted only once and can be used for obtaining solutions for a total set of plane waves with wave vectors $\vec{k}$ associated with equally located points in different Brillouin zones. Such sets of wave vectors are called Bloch families. In Fig. 2.8 this is visualised. The dots inside the Brillouin zones are of equal colour for wave vectors of the same Bloch family. Additionally, the circle related to the cutoff wave vector $\vec{k}_c$ is illustrated. The finite element system matrix, which is necessary to calculate the solution, is identical for wave vectors belonging to the same Bloch family. Since the inversion of the system matrix is the most time-consuming part of the simulation, this reduction of complexity saves a considerable amount of computational time [167]. For sampling of the irreducible Brillouin zone, a Monkhurst-Pack grid is chosen for which the sampling points are equidistantly distributed along the reciprocal lattice vectors [168]. What in the end needs to be done is to use a multi-layer solver for the final equations connected by the scattering matrices. Examples of such multi-layer solvers can be found in literature [169–171]. In the case of this work a set of linear equations is created to connect the incoming and outgoing flux at the interfaces. By solving this set of equations the incoming and outgoing fluxes at each interface can be determined [172]. In that way the desired value of the transmittance from above through the nanopatterned AlN/sapphire interface can be simulated for all relevant modes. The total transmittance is then determined by numerical integration over the whole angular distribution normalized to the area of the hemisphere.

CHAPTER 3

High temperature annealing of MOVPE grown AlN

The method of HTA in face-to-face configuration was firstly demonstrated by the group of Prof. Miyake in 2016 at the Mie university in Japan for sputtered AlN templates. The outstanding improvement of the material quality by this relatively simple approach (details described in chapter 1.3) had a great impact on the UV LED community.

In this chapter, HTA in face-to-face configuration will be applied to MOVPE grown AlN templates. The aim of applying this technique is to improve the material quality of the MOVPE AlN by reducing the TDD analogous to sputtered AlN. While a low TDD is of crucial importance for an AlN template, other criteria, e.g. the surface roughness, the strain of the layer and the curvature of the wafer also have to be taken into account. A comprehensive study on the characteristics of the AlN layers before and after annealing as well as annealing parameters such as temperature and time will be presented and evaluated regarding the named criteria. The purpose is to develop a reproducible technology while considering, discussing and trying to solve possible drawbacks along the process chain from the AlN template to the final UV LED device.

In the first section of this chapter the influence of the AlN properties, e.g. strain and layer thickness upon the outcome after annealing will be investigated and discussed. Based on this analysis, in the second section, the ideal characteristics for the AlN layers used in further annealing experiments are extracted. The impact of varying annealing parameters, e.g. temperature and time will be examined in the third section. A short resume with the most important findings and the conclusion can be found at the end of the chapter.

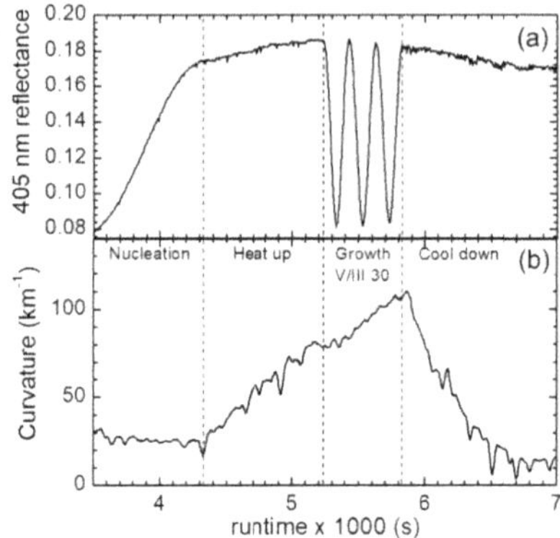

Figure 3.1: (a) 405 nm reflectance and **(b)** curvature measured in-situ during growth of the 350 nm thick AlN layer. The Fabry-Pérot oscillations of the reflectance suggest smooth growth, while the increasing curvature indicates the build-up of tensile strain. From reference [173] (first authorship) with permission of The Japan Society of Applied Physics © 2019.

3.1 Annealing of 350 nm thick AlN

As starting point for the face-to-face annealing experiments, a thickness of 350 nm AlN is chosen, since this layer thickness showed good results for annealing sputtered AlN layers in face-to-face configuration [105]. 2-inch sapphire with *c*-orientation and offcut of 0.2° towards the *m* direction is used as substrate. Prior to the actual AlN growth, organic contamination on the sapphire is removed by heating up the MOVPE reactor under hydrogen flow. This thermal cleaning step was performed at a process temperature of $T_{\mathrm{Proc}} = 1240\,°\mathrm{C}$ for 10 min. After thermal cleaning the reactor was cooled down to $T_{\mathrm{Proc}} = 980\,°\mathrm{C}$ to grow the 50 nm thick nucleation layer at a V/III ratio of 4000 for 15 min [55]. In the following parts of this work, the thermal cleaning and nucleation stay unchanged and are referred to as standard thermal cleaning and standard nucleation. After nucleation the reactor was heated up to $T_{\mathrm{Proc}} = 1380\,°\mathrm{C}$ for AlN growth at V/III ratio of 30 for 10 min resulting in growth of 300 nm AlN (growth rate of $1.6\,\mathrm{\mu m\,h^{-1}}$). Together with the 50 nm nucleation layer this yields 350 nm thick AlN. Finally, the reactor was cooled down to room temperature.

The growth was monitored in-situ including reflectance of the wafer surface at 405 nm wavelength and curvature of the wafer. These parameters are plotted against the runtime in Fig. 3.1 from nucleation until cool-down. During nucleation the 405 nm reflectance increases from 0.08 up to 0.18. This is related to the higher reflectance of

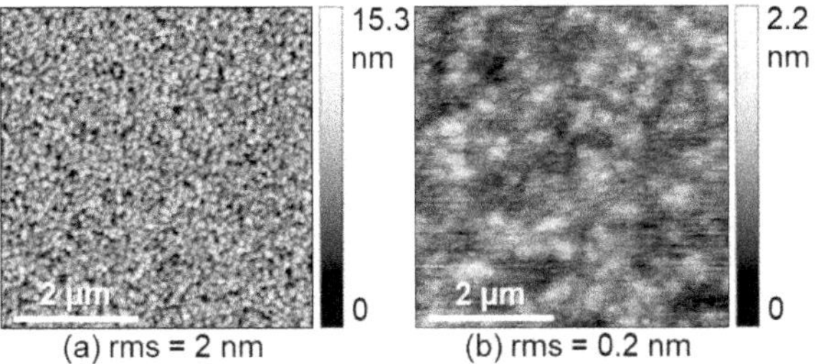

Figure 3.2: AFM images of **(a)** the 50 nm standard nucleation layer and **(b)** the 350 nm thick AlN layer. After full coalescence of the nucleation grains the 350 nm thick AlN layer shows an atomically smooth surface with the typically observed AlN bilayer steps and an rms value as low as 0.2 nm for an area of 5 µm x 5 µm.

AlN compared to sapphire due to their different refractive indices. Another reason is the formation of Fabry-Pérot oscillations due to the increasing AlN layer thickness. The period of one full oscillation in case of the refractive index profile of AlN/sapphire and a wavelength of 405 nm is approximately 100 nm. Hence, after the 50 nm nucleation layer, the peak-to-peak amplitude of the first oscillation is reached. The curvature stays approximately constant during nucleation. An AFM image of the standard nucleation layer is depicted in Fig. 3.2a: Small AlN grains of 50 nm width in columnar arrangement can be observed. They do not form a continuous 2D layer yet and thus, there is no remarkable strain impact affecting the curvature of the wafer. Heating up the reactor to growth temperature leads to a slight increase of the reflectance due to the temperature dependence of the refractive index. The curvature is strongly increasing caused by the thermal mismatch between AlN and sapphire and the temperature gradient between the heated backside of the wafer and the wafer's surface. As soon as the AlN continues to grow the reflectance shows additional Fabry-Pérot oscillations. The amplitude, period and average value of the oscillations stay constant for the whole growth which indicates smooth and uniform growth. The curvature of the wafer reaches a short period of being constant after heating up, but soon continues to rise as the AlN grows. Considering the surface morphology after growth measured by AFM (Fig. 3.2b) it can be seen that the columnar nucleation grains developed into a continuous 2D layer. This happens shortly after nucleation since the relatively high growth temperature of $T_{\mathrm{Proc}} = 1380\,°\mathrm{C}$ and low V/III ratio of 30 enhances the diffusion length of the AlN adatoms promoting lateral growth and thus, resulting in fast coalescence of the nucleation grains. After

Table 3.1: XRC-FWHM of 0002 and 10-12 reflections of the 350 nm thick AlN layer before and after annealing at 1700 °C for 3 hours. The TDDs are estimated by calculation from the XRC FWHM and show a reduction of more than one order of magnitude due to annealing.

	XRC-FWHM of 0002	XRC FWHM of 10-12	Estimated TDD
As grown	100″	1700″	$3 \times 10^{10}\,\mathrm{cm}^{-2}$
After annealing	50″	450″	$2 \times 10^{9}\,\mathrm{cm}^{-2}$

full coalescence of the nucleation grains, the curvature increases due to the build-up of tensile strain which is usually observed for AlN growth on sapphire. During cool down the reflectance shrinks according to the temperature dependence of the refractive index. The curvature decreases until it reaches the final value of $10\,\mathrm{km}^{-1}$ determined by the thermal mismatch between AlN and sapphire as well as the strain in the AlN layer at growth temperature. The full growth including nucleation results in a layer thickness of approx. 350 nm This was determined by a fit of the 405 nm reflectance curve using the software EpiNet from LayTec. In the AFM image of the final layer (Fig. 3.2b) the rms value is 0.2 nm (for an area of 5 µm x 5 µm) and taking a closer look, the typical AlN bilayer steps can be spotted indicating an atomically smooth surface. The AlN bilayer steps have a height of approx. 0.25 nm which is the height of the AlN wurtzite unit cell. The material quality was examined by XRD: XRC-FWHM of the 0002 and 10-12 reflections were measured and are listed in Tab. 3.1. The XRC-FWHM of the 0002 reflection is a measure for the misalignment of the AlN grains in c-direction (tilt), while the XRC-FWHM of the 10-12 reflection is a measure for the rotational misalignment (twist) with the c-direction as the rotational axis. The misalignment is compensated by TDs. Tilt leads to screw and twist to edge TDs. The XRC-FWHM of the 10-12 reflection of 1700″ is much higher than the XRC-FWHM of the 0002 reflection of 100″. Hence, edge-type dislocations are the dominant type of TDs in the layer. An overall TDD can be estimated by calculation from the XRC-FWHM of the 0002 and 10-12 reflections [142, 147]. Details of this procedure are described in chapter 2.2. The estimated TDD for the 350 nm thick AlN layer was calculated to $3 \times 10^{10}\,\mathrm{cm}^{-2}$.

The 350 nm thick AlN layer was then annealed at 1700 °C for 3 hours in the annealing furnace presented in chapter 2.1 located at the Berlin company HTM Reetz. After annealing, the XRC-FWHM were measured again (Tab. 3.1). The XRC-FWHM of the 0002 reflection shows a reduction from 100″ to 50″, while for the 10-12 reflection the XRC-FWHM reduces from 1700″ down to 450″. This results in a reduction of the overall TDD by more than one order of magnitude from $3 \times 10^{10}\,\mathrm{cm}^{-2}$ to $2 \times 10^{9}\,\mathrm{cm}^{-2}$,

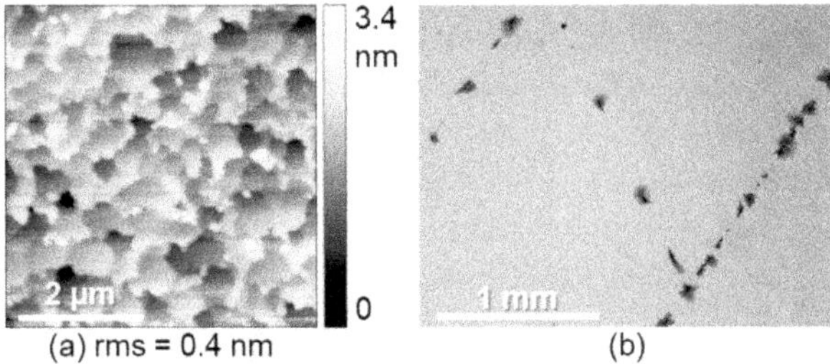

Figure 3.3: (a) AFM image and **(b)** light microscopy image with Nomarski contrast of the 350 nm thick AlN layer after HTA at 1700 °C for 3 hours. The microscopic surface morphology consists of small step bunches of approx. 1 nm height, while the macroscopic surface shows cracks. **(b)** from reference [173] (first authorship) with permission of The Japan Society of Applied Physics © 2019.

which can be explained following the model of Washiyama et al. [129]: The elevated temperature of 1700 °C enables the TDs to move by dislocation climbing inside of the AlN layer so that they can react with each other. If dislocations with opposite Burgers vector meet, they annihilate and hence, the overall TDD is reduced [57]. However, this only happens if their radius of movement is big enough to be attracted by the strain field of another dislocation with opposite Burgers vector. The radius of movement is limited by the energy of the dislocations which in this case is provided by the high temperature during annealing. Since dislocations are non-equilibrium crystal defects, they tend to annihilate if enough energy is supplied so that the system develops towards thermodynamic equilibrium. Such recovery effects are common in metals and were also observed for other materials, e.g. the mineral Calcite ($CaCO_3$) [174]. To investigate the influence of HTA on the microscopic surface morphology of the AlN layer, AFM images were acquired and can be seen in Fig. 3.3a. The surface gets rougher after annealing compared to the as grown layer (Fig. 3.2b) resulting in an increase of the rms value from 0.2 nm to 0.4 nm (for an area of 5 µm x 5 µm). Small step bunches with height of approx. 1 nm were formed due to annealing so that the surface is not atomically smooth anymore. This phenomenon can be explained by mass transport at the surface during annealing: The temperature is high enough for decomposition of small parts of AlN leading to a constant exchange of desorbing and adsorbing Al and N adatoms analogous to usual epitaxial growth. The diffusion length of the adsorbed Al and N adatoms is relatively long due to the high temperature leading to the build-up of step bunches analogous to epitaxial growth in step bunching growth mode [175, 176]. Step bunches of only 1 nm height are unproblematic, since they can easily be smoothed by additional MOVPE growth in step-flow growth mode. The macroscopic surface morphology was examined using light microscopy imaging with Nomarski contrast (Fig. 3.3b). Cracks penetrating over the whole 2-inch wafer can be seen. The cracking is ascribed to the annealing process since the layer was checked before annealing and did not contain any cracks. An explanation for this cracking could be the high amount of tensile strain during growth inside the layer already before annealing indicated by the strong increase of the curvature during growth (Fig. 3.2b). The thermal mismatch between AlN and sapphire results in larger expansion of the sapphire compared to the AlN layer during annealing at temperatures higher than the growth temperature. This further increases the tensile strain until it partially relaxes via generation of macroscopic cracks. Cracking of layers caused by HTA was also observed by Zhao et al. for sputtered AlN layers of 1000 nm containing a large amount of tensile strain already before annealing similar to the cracks observed in the 350 nm thick AlN layer after HTA [177].

While the microscopic surface of the layer after annealing looks promising, the cracking makes the annealed 350 nm thick AlN layer unsuitable to work as a base layer for UV LEDs although the TDD was reduced by more than one order of magnitude due to HTA. To avoid layer cracking during HTA, lower tensile strain in the initial AlN layer could be an option. The preparation of such layers and subsequent investigations regarding HTA is the topic of the next section.

3.2 Annealing of layer thickness series

To overcome the issue of layer cracking during annealing, a growth method suppressing the build-up of tensile strain has to be found. This in general can be induced by gradual relaxation via certain relaxation channels in order to avoid the continuous build-up of tensile strain. An interesting approach to achieve this is to provoke roughening and subsequent smoothing of the AlN layer by applying appropriate growth conditions [84]. Besides the general reduction of tensile strain during growth, this approach offers crack-free growth of thick AlN layers.

The method of roughening and subsequent smoothing was applied as follows: After standard thermal cleaning and standard nucleation the reactor was heated up to the growth temperature of $T_{\mathrm{Proc}} = 1180\,^{\circ}\mathrm{C}$ which was kept constant until cool down. The V/III ratio was first set to 540 for 17 min (growth rate $1.6\,\mu\mathrm{m\,h^{-1}}$) before it was gradually decreased down to 30 over a time of 15 min. The final V/III ratio of 30 was kept constant

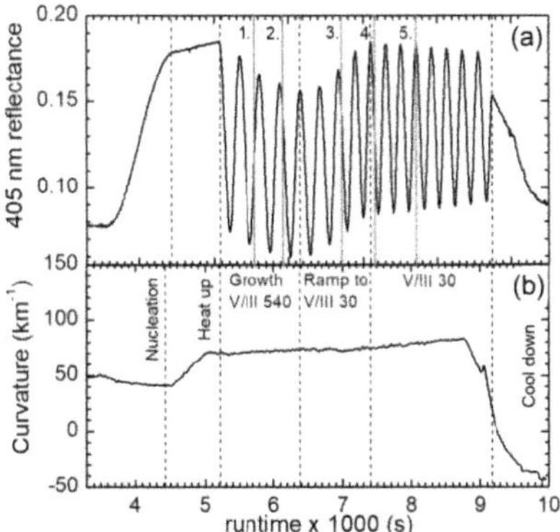

Figure 3.4: (a) 405 nm reflectance and **(b)** curvature measured in-situ during growth of 1.5 µm AlN. Roughening and subsequent smoothing of the layer prevents the build-up of tensile strain so that 1.5 µm crack-free AlN on sapphire is achieved. The growth process is interrupted at the 5 marked positions for subsequent HTA. From reference [173] (first authorship) with permission of The Japan Society of Applied Physics © 2019.

for another 30 min until the reactor was cooled down to room temperature. The change of the growth rate due to the different V/III ratios is negligible. The in-situ measurement of the growth process including the 405 nm reflectance and curvature of the wafer is depicted in Fig. 3.4. The behaviour under nucleation, heating up and cooling down is the same as shown previously (Fig. 3.1a). At the initial V/III ratio of 540 the average reflectance of the Fabry-Pérot oscillations decreases from 0.13 to 0.1 indicating roughening of the AlN layer. During the ramp to the V/III ratio of 30 the average reflectance starts to increase again until 0.13 is reached which indicates a smooth AlN layer. From then on the average reflectance stays constant at 0.13. The curvature of the wafer during roughening only slightly increases and stays almost constant. As soon as the average reflectance of 0.13 is reached, the slope of the curvature increases. However, if compared to the 300 nm thick layer grown before (Fig. 3.1b), the overall increase of the curvature is much less. So from the in-situ measurement it can be concluded that the desired suppression of the build-up of tensile strain was achieved.

In order to better understand this growth process, 5 samples were prepared interrupted at different positions in the growth process (as marked in Fig. 3.4a) for acquiring AFM images which are depicted in Fig. 3.5a.1-a.5. At position 1 (Fig. 3.5a.1) the layer consists of many holes. The formation of these holes is the reason for the decrease of the average reflectance in Fig. 3.4a. The rise of the rms value from 0.2 nm for the AlN

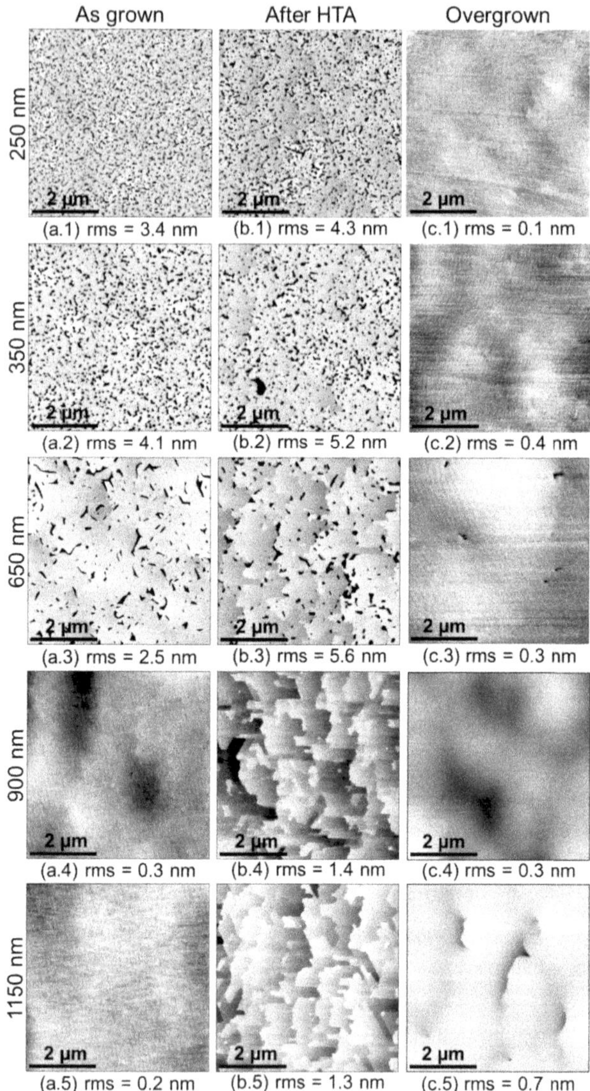

Figure 3.5: AFM images of 250 nm, 350 nm, 650 nm, 900 nm and 1150 nm thick AlN as grown **(a.1-5)**, after HTA at 1700 °C for 3 hours **(b.1-5)** and after overgrowth up to a total layer thickness of 1500 nm **(c.1-5)**. From reference [173] (first authorship) with permission of The Japan Society of Applied Physics © 2019.

nucleation layer (Fig. 3.2) to 3.4 nm (for an area of 5 µm x 5 µm) quantitatively shows this roughening. The roughening continues from position 1 to position 2 noticeably by the increase of the hole size (Fig.3.5a.2) and a higher rms value of 4.1 nm (for an area of 5 µm x 5 µm). According to the reflectance measurement, the smoothing of the layer starts at position 3 which is in good agreement with the corresponding AFM image (Fig. 3.5a.3): Shrunk holes and a reduced rms value of 2.5 nm (for an area of 5 µm x 5 µm) can be observed due to growth pointing to coalescence of the layer. The process of coalescence resulting in the smoothing is finished at positions 4 and 5 (Fig. 3.5a.4+a.5) for which an atomically smooth layer in step-flow growth mode is reached. The correlated rms values are as low as 0.3 nm and 0.2 nm (for an area of 5 µm x 5 µm). The model behind this roughening/smoothing process is not straightforward to understand. A detailed explanation is given in the following. The standard nucleation of the AlN starts with a simultaneous flow of ammonia and TMAl. This leads to a partial nitridation at growth start inducing mixed polarity growth of Al-polar and N-polar domains [178]. Depending on the growth conditions this arrangement develops differently: For the 300 nm thick layer in the previous section of this chapter, the high growth temperature of 1380 °C and low V/III ratio of 30 strongly favours lateral growth of the Al-polar nucleation islands leading to fast overgrowth of the N-polar islands and coalescence of the Al-polar domains. This process can be retarded by changing the growth conditions according to the described roughening/subsequent smoothing process. The lower growth temperature of 1180 °C and high V/III ratio of 540 promotes growth of the Al-polar nucleation islands in *c*-direction. In contrast to lateral growth the fast growth in *c*-direction leads to a reduced area of the *c*-facet. This explains the formation of holes (positions of the N-polar domains) and their tendency to increase in size. By decreasing the V/III ratio to 30, lateral growth is favoured resulting in an increasing area of the *c*-facet. This happens until the Al-polar nucleation grains start to coalesce forming a continuous 2D layer. As long as the layer is porous before coalescence of the nucleation grains at positions 1 and 2 in Fig. 3.4, there is no build-up of tensile strain. After coalescence of larger areas at position 3 in Fig. 3.4b, the slope of the curvature increases indicating that after coalescence tensile strain starts to be incorporated. However, the overall build-up of tensile strain is reduced efficiently so that this growth process enables crack-free AlN on sapphire on the full 2-inch wafer up to a thickness of approx. 1.5 µm. Consequently, this growth method opens up the possibility of investigating HTA of MOVPE AlN of different thickness with reduced tensile strain as was the desired outcome of the last section of this chapter.

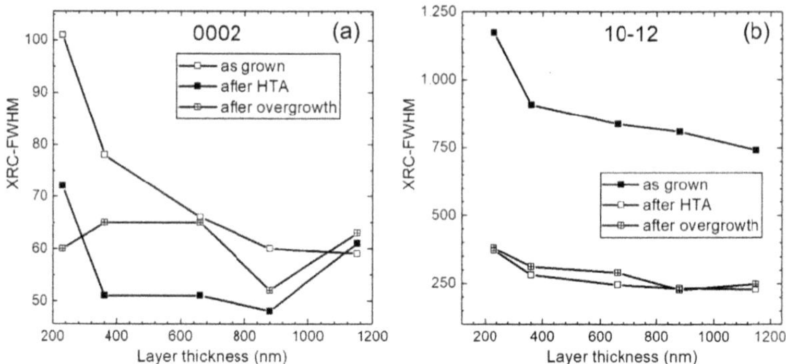

Figure 3.6: XRC-FWHM of **(a)** 0002 and **(b)** 10-12 reflections against layer thickness as grown, after HTA at 1700 °C for 3 h and after overgrowth up to a total layer thickness of 1.5 μm. The broadening of the 10-12 reflection related to edge-type TDs is the dominant broadening for all layers and shows the lowest values for the thicker layers for all cases as grown, after HTA and after overgrowth.

For this purpose, the 5 different layers with thickness between 250 nm and 1150 nm according to the marked positions in Fig. 3.4a prepared for the AFM measurements were used. They were annealed at 1700 °C for 3 hours in the annealing furnace at HTM Reetz analogous to the procedure described in the previous section of this chapter. The first result is that after HTA the layers did not show any cracks. This was checked by light microscopy imaging. As a conclusion, the preparation and annealing of layers with less incorporated tensile strain met the expectation of suppressing the formation of macroscopic cracks on the surface of the layers during HTA. The material quality was determined by measuring the XRC-FWHM of the 0002 and 10-12 reflections for the as grown layers and after HTA (Fig. 3.6). Before annealing, a decrease in the XRC-FWHM with increasing layer thickness for both reflections is observed. This can be explained by the mutual annihilation of TDs with increasing layer thickness during growth similar to other material systems [146, 179]. Since not all TDs are perfectly parallel to the growth direction, they have a certain probability to meet and react leading to annihilation for TDs with opposite Burgers vectors. This effect gets less pronounced for lower TDDs when the average distance of the TDs becomes too high, which happens for the thick layers. Similar to the 300 nm thick layer in the previous section of this chapter the 10-12 reflections are much broader than the 0002 reflections meaning edge-type TDs are the dominant TDs in all the layers. After HTA the XRC-FWHM of the 0002 reflection vary only slightly (Fig. 3.6a), while all the layers exhibit a distinct reduction of the

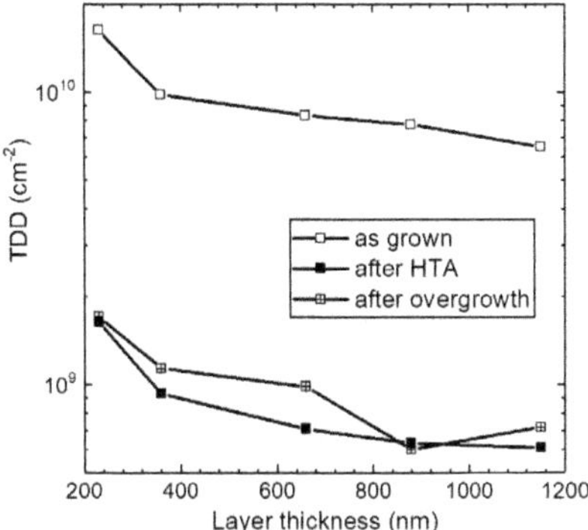

Figure 3.7: TDDs estimated by calculation from the XRC-FWHM shown in Fig. 3.6 against layer thickness as grown, after HTA at 1700 °C for 3 h and after overgrowth up to a total layer thickness of 1.5 µm. The lowest values are achieved for the thicker layers in all cases as grown, after HTA and after overgrowth.

XRC-FWHM of the 10-12 reflection (Fig. 3.6b). Especially for the thick layers, the density of the TDs with screw type component is already too low for another significant reduction in contrast to the TDs with edge-type component. Another point to remark is that for the XRC-FWHM of the 10-12 reflection, there is still a relation between the initial and resulting value after HTA so that the overall trend of the XRC-FWHM is the same as for the as grown layers and the lowest values are reached for the thick layers with the lowest initial values (900 nm and 1150 nm). The TDDs estimated by calculation from the XRC-FWHM are plotted in Fig. 3.7. Obviously, the TDDs follow the same trend as the XRC-FWHM of the 10-12 reflection. For the as grown layers, the TDDs are in the range of $1 \times 10^{10}\,\mathrm{cm}^{-2}$, while after HTA values below $1 \times 10^{9}\,\mathrm{cm}^{-2}$ were achieved. So the reduction of one order of magnitude observed in the previous section of this chapter for the 300 nm thick layer is valid for the whole thickness series. The lowest TDDs are as low as $6 \times 10^{8}\,\mathrm{cm}^{-2}$ for both the 900 nm and 1150 nm thick layers.

The surface morphology after HTA was checked by AFM. The related images are depicted in Fig. 3.5b.1-5. Before coalescence (Fig. 3.5b.1-3), the holes of the layers become slightly bigger during annealing. Additionally, for all layers (Fig. 3.5b.1-5) a redistribution of the surface takes place so that the formerly bilayer steps form small step bunches similar to the result of the 300 nm thick layer in the previous section of this chapter. This can be seen most clearly for the already coalesced layers of 900 nm and 1150 nm thickness (Fig. 3.5b.4-5). Due to the increase of the hole size and the formation of step bunches, all layers get rougher which translates into higher rms values. The explanation is the same as for the 300 nm thick layer in the previous section of this chapter: The AlN adatoms on the surface become mobile with a large diffusion length due to the elevated temperature and hence, formation of step bunches similar to epitaxial growth in step bunching growth mode occurs. The increase in the diameter of the holes that are still present after HTA in the layers of 250 nm to 650 nm thickness (Fig. 3.5b.1-3) can be explained by redistribution and desorption of material from the hole edges resulting in formation of more extended stable AlN sidewall facets.

To gain a deeper insight of the impact of HTA upon the morphological characteristics of the AlN layers, TEM investigations were conducted for the 650 nm thick layer before and after annealing. The TEM images acquired by Leonardo Cancellara at the Leibniz-Institut für Kristallzüchtung (IKZ) located in Berlin are shown in Fig. 3.8. In Fig. 3.8a+d cross-section STEM-HAADF images are shown for the as grown and annealed 650 nm thick AlN layers. At growth start of the as grown layer, a lot of contrast can be seen due to many TDs and the mixed polarity domains in the nucleation layer. This contrast decreases with increasing layer thickness, since the film becomes purely Al-polar and also the TDs reduce due to mutual annihilation. Additionally, the voids penetrating through the layer from growth start up to the surface are present. In the image after annealing, the whole contrast arrangement changes remarkably. At the AlN/sapphire interface only a single sharp white line remains. This line was more closely investigated by HR-TEM and identified to be the inversion domain boundary between N-polar AlN and Al-polar AlN similar to other reports in literature [102, 180]. Furthermore, a lot of the dislocation related contrast vanishes due to the efficient reduction of the TDD during annealing. The voids are still present. However, some of them appear shaped like bubbles with brighter contrast compared to the surrounding material. The material with bright contrast inside of these bubbles was more closely investigated by HR-TEM and ascribed to γ-AlON. By taking the phase diagram of AlN and sapphire at elevated temperatures in the work of Bandyopadhyay et al. into account, it is found that the formation of AlON becomes probable at elevated temperatures [181]. The oxygen needed for that is most likely supplied by diffusion from the sapphire through the voids. In Fig. 3.8b+e,c+f the diffraction contrast for beam conditions with <0002> and <11-20> g-vectors are shown for the as grown and annealed layer, respectively. Under a beam condition of <0002>, only dislocations with a screw component are visible. While for the as grown layer some dislocations with screw component can be observed, for the annealed layer no dislocations with screw component could be observed in the related

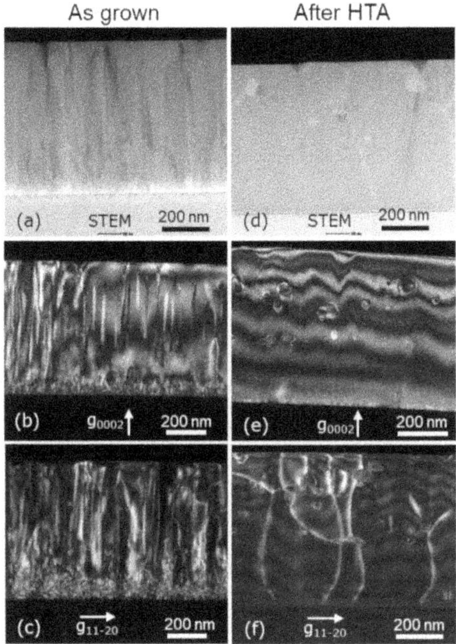

Figure 3.8: (a)+(d) Cross-sectional STEM-HAADF images and cross-sectional TEM-diffraction contrast images taken under **(b)+(e)** 0002 diffraction condition and **(c)+(f)** 11-20 diffraction condition of 650 nm thick AlN layer as grown and after HTA at 1700 °C for 3 h, respectively. The reduction of the TDD in the layer after HTA is clearly visible by the reduced dislocation contrast in all images. Besides, the formation of γ-AlON was identified by HR-TEM inside some of the bubble-like structures. With STEM-HAADF it exhibits brighter contrast compared to AlN. The TEM investigations were carried out by Leonardo Cancellara at IKZ located in Berlin.

image of the specimen. For the <11-20> beam condition, only dislocations with an edge component are visible. A massive reduction of the visible dislocations after annealing takes place similar to the <0002> beam condition. In this case, however, there are still some dislocations with edge component left in the related image of the specimen. The TDD can be extracted from the acquired images by determining the volume of the specimen considering its thickness and then simple counting of the visible TDs. For the as grown layer, this method yields a TDD of $1.4 \times 10^{10}\,\mathrm{cm}^{-2}$ and $1.2 \times 10^{9}\,\mathrm{cm}^{-2}$ for the annealed layer. These values are slightly higher than the values calculated from the XRC-FWHM (Fig. 3.7) but show the same trend of the reduction of the TDD by

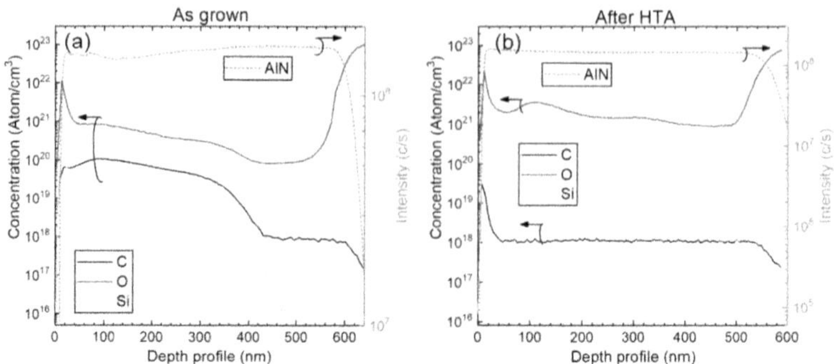

Figure 3.9: Depth profiles of the impurity concentrations of O, Si and C measured by SIMS for the 650 nm thick AlN layer **(a)** as grown and **(b)** after HTA. The depth profiles indicate diffusion in the AlN/sapphire layer stack. The peak in the O concentration about 100 nm beneath the surface after HTA is related to formation of AlON induced by oxygen diffusion from the sapphire into the AlN. The measurement was carried out at the company RTG located in Berlin.

one order of magnitude due to HTA.

The formation of AlON inside the 650 nm thick AlN layer after HTA revealed by the TEM investigations is ascribed to diffusion of oxygen during annealing. To further investigate the diffusion processes taking place inside of the layer during annealing, impurity concentrations inside of the 650 nm thick AlN layer as grown and after annealing were determined by SIMS measurements. This was done for the impurities oxygen (O), silicon (Si) and carbon (C) by the company RTG located in Berlin. The results are shown in Fig. 3.9 for the 650 nm thick AlN layer as grown and after HTA. Additional to the impurity concentrations O, Si and C, the signal of AlN was measured to identify the AlN/sapphire interface. The depth profiles start at the surface of the layer penetrating through it until the sapphire is reached at the point for which the AlN signal decreases and the oxygen concentration strongly increases. One thing to keep in mind is that the layer is not fully coalesced and so the actual depth profiles of the SIMS measurement might be distorted by reason of signal coming not only from the surface but also through the voids from different depths of the layer. For the as grown layer, all impurity concentrations show the lowest value for the first approx. 50 nm to 100 nm of AlN after the sapphire. This part mainly consists of the nucleation layer. After the nucleation layer when the hole-like surface starts to build-up, also the impurity concentrations rise. This is most likely due to the increased surface area emerging due to the fast growth of the Al-polar nucleation islands during the roughening part of the growth process. During the rough growth in which no continuous 2D layer is achieved, impurities from the reactor atmosphere are built in on the sidewalls of the Al-polar nucleation islands over a long period of time, since the lateral growth rate is very small. Impurity concentrations of $1 \times 10^{21}\,\mathrm{cm}^{-3}$ for oxygen, $5 \times 10^{19}\,\mathrm{cm}^{-3}$ for silicon and $1 \times 10^{20}\,\mathrm{cm}^{-3}$ for carbon were observed. After HTA the depth profiles of Si and C are strongly flattened while the overall concentration seems to reduce. This indicates diffusion inside the whole AlN layer and eventually further into the sapphire so that in thermodynamic equilibrium the initial concentration is uniformly distributed. Diffusion into the sapphire would explain the reduction of the overall concentration of Si and C in the AlN. Another possibility is that part of the impurities left the sample through its free surfaces. The oxygen profile behaves differently showing an overall increase in the concentration and a small peak at around 100 nm beneath the surface. This peak with a concentration of approx. $3.5 \times 10^{21}\,\mathrm{cm}^{-3}$ can be correlated to the AlON which formed inside of the voids as identified in the TEM images in Fig. 3.8d. The additional oxygen supply can be attributed to oxygen diffusion from the sapphire substrate into the AlN layer. This was already assumed from the TEM images, since enough oxygen needs to be supplied for the formation of AlON inside of the bubbles.

The final concentrations of C and Si being in the range of $10^{18}\,\mathrm{cm}^{-3}$ are not an issue for further LED growth. However, a high concentration of oxygen inside the AlN can lead to defect complexes which can lead to absorption in the UV range [182, 183]. If the absorption lies in the same region as the wavelength of the final LED this obviously reduces the output power. To check eventual absorption in the relevant UV range, the transmittance of the 650 nm thick AlN layer as grown and after HTA was measured. The transmittance is plotted in Fig. 3.10 for wavelengths between 200 nm and 1100 nm. The first thing to note is that the annealed layer in general shows a higher transmittance than the as grown layer. This can be explained by the following: Since the usual sapphire substrates are single-side polished only on the surface to grow on, a lot of the light is scattered at the rough backside of the sapphire. This light can not be collected in the used measurement setup and is therefore not contributing to the measured transmittance. During HTA the sapphire backside is smoothed leading to an enhanced transmittance, which can be seen with the bare eye for the visible spectral range. The transmittance measurement shows that this is also valid for the UV spectral range. During the final LED chip process, the backside of the sapphire is anyway treated so that there will not be a difference for the final device. Hence, this offset can be neglected in the following discussion. Above the bandgap of AlN at a wavelength

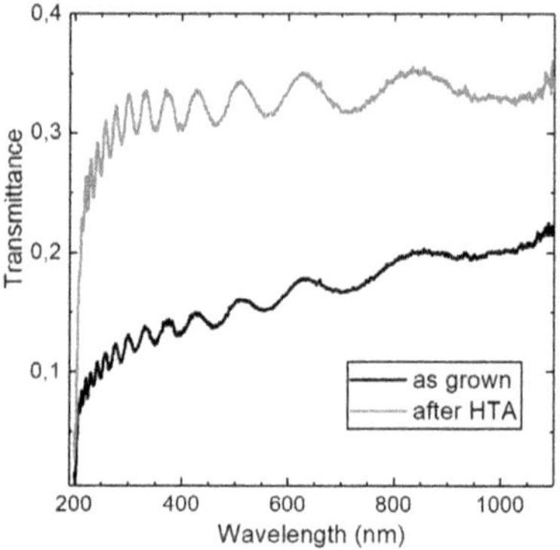

Figure 3.10: Transmittance of the 650 nm thick AlN layer as grown and after annealing. The increased transmittance after HTA is ascribed to smoothing of the rough backside of the sapphire during annealing. No absorption is detected for the layer after HTA.

of 206 nm the expected drop of the transmittance is observed. Above the band edge Fabry-Pérot oscillations caused by the layer stack of 650 nm thick AlN on sapphire can be seen. The average transmittance is increasing for larger wavelengths. At the most popular wavelength for UV LEDs of 265 nm, the average transmittance is 0.11 for the as grown and 0.3 for the annealed sample. What is important to notice is that no local minimum of the transmittance related to absorption was found for the layer after HTA. So although the oxygen concentration inside of the layer is relatively high, this does not lead to a distinct decrease in the transmittance. Hence, absorption of the emitted light in the final LED structure in the annealed AlN should be negligible.

After the extended structural characterisation of the annealed samples, the growth process shown in Fig. 3.4 was continued for all samples up to a total AlN layer thickness of 1.5 µm. This was done to see the impact of the intermediate annealing step introduced at different positions of the full growth process. Exemplarily, an in-situ measurement of overgrowing the 900 nm thick AlN layer with intermediate HTA is shown in Fig. 3.11a+b. For comparison, a simultaneously grown 900 nm thick AlN layer without intermediate HTA was overgrown in the same growth run. For both samples the 405 nm reflectance (Fig. 3.11a) behaves identical: Fabry-Pérot oscillations with an average reflectance of 0.13 can be seen which indicate growth of a smooth AlN layer. This identical behaviour regarding the 405 nm reflectance is the same for overgrowing all layers of different thickness. So the first conclusion is that there is no general disturbance of the surface for overgrowing the layers which were annealed at a high temperature

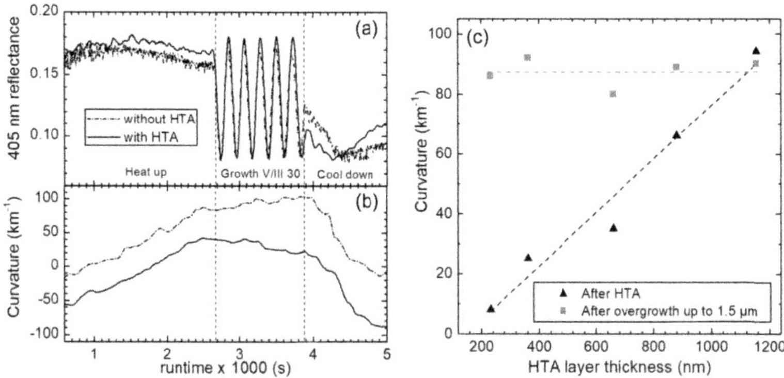

Figure 3.11: (a) 405 nm reflectance and **(b)** curvature measured in-situ during overgrowth of the 900 nm thick AlN layer with and without intermediate HTA up to the total layer thickness of 1.5 µm. **(c)** Curvature against layer thickness after HTA and after overgrowth up to the final layer thickness of 1.5 µm. The annealed layers grow compressive in contrast to the normally observed tensile growth. This change of the strain translates to larger wafer bow, which for the fully overgrown 1.5 µm thick AlN layers is independent of the layer thickness at the intermediate HTA. From reference [173] (first authorship) with permission of The Japan Society of Applied Physics © 2019.

and that the growth develops in principle equally with and without the intermediate HTA. The curvature measurement (Fig. 3.11b), however, shows a different behaviour for the two layers with and without the intermediate HTA. Firstly, before heating up, the sample with HTA exhibits a larger wafer curvature of $-60\,\text{km}^{-1}$ in comparison to $-10\,\text{km}^{-1}$ for the sample without HTA. Secondly, for the layer with intermediate HTA, the curvature value decreases during growth indicating build-up of compressive strain rather than the build-up of tensile strain observed for the sample without HTA. This observation is in agreement with previous studies of high temperature annealed samples which showed smaller in-plane lattice constants after annealing [105,108] due to relaxation of the AlN layer at the high annealing temperature [128]. The smaller in-plane lattice constant leads to compressive growth under the applied growth conditions. The wafer curvature after HTA linearly increases with layer thickness (Fig. 3.11c) reaching a constant value of about $-90\,\text{km}^{-1}$ after overgrowth up to the total layer thickness of 1.5 µm for all samples. This is the expected behaviour assuming a reduced in-plane lattice constant after annealing which is independent of the starting layer thickness. By looking at the XRC-FWHM of the overgrown samples, it can be seen that they tend to slightly increase. This can be ascribed to the generation of new TDs during the compressive growth. In this case further reduction with increasing the layer thickness is not observed, since for that mechanism the TDD is already too low. However, the final TDD (Fig. 3.7) still shows good results especially for the two layers of large thickness (900 nm and 1150 nm) with final TDDs in the range of 6-7 $\times 10^8$ cm^{-2}.

The AFM images after overgrowth (Fig. 3.5 c.1-5) clearly show that in all cases step-flow growth was achieved with mostly atomically smooth surfaces. However, for the thickest layer (Fig. 3.5 c.5), remnants of step bunches generated during HTA can be seen. Hence, the growth of additional 350 nm is not sufficient to properly smoothen the step bunches. For 900 nm the additional growth of 600 nm is sufficient to reach an atomically smooth surface again. For the 650 nm thick layer, some holes can be observed surrounded by an atomically smooth surface. This can be explained by the increase of the hole diameters and the formation of more stable AlN facets during annealing so that some of the holes could not be overgrown properly. For the thinner layers no holes are observed after overgrowth, since in that case the holes were smaller.

So with respect to surface smoothness, the overgrown layers with starting thickness of 250 nm, 350 nm and 900 nm all show the desired atomical smoothness for UV LED base layers. However, the TDD of the two thinner starting layers is still relatively high in contrast to the low value of 6×10^8 cm^{-2} for the 900 nm starting layer. The wafer curvature after overgrowth is independent of the thickness of the starting layer. So to conclude which layer works best regarding TDD, surface smoothness and wafer bow, the 900 nm starting layer is chosen and will be used for the following annealing experiments. In the next section of this chapter, the variation of the annealing parameters will be investigated to further improve the AlN material quality.

3.3 Variation of the annealing parameters

For the following annealing experiments, nominally equal AlN layers of 900 nm thickness prepared as described in the previous section of this chapter are used. The main parameters for annealing are the annealing temperature and the annealing time. The atmosphere (nitrogen) and pressure (atmospheric pressure) will not be changed. The following annealing experiments were conducted in the annealing furnace located at FBH. Details of this annealing furnace are described in chapter 2.1. The temperature was varied between 1300 °C and 1710 °C at a constant annealing time of 1 h. The time series consists of annealing temperatures of 1600 °C and 1710 °C while the annealing time was varied between 1 h and 5 h.

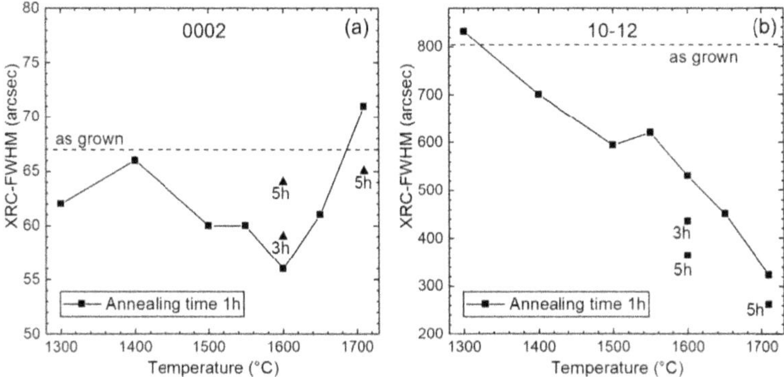

Figure 3.12: XRC-FWHM of **(a)** 0002 and **(b)** 10-12 reflections for 900 nm thick AlN layers annealed at different temperatures and for different time spans. While the XRC-FWHM of the 0002 reflection do not show a clear trend, the XRC-FWHM of the 10-12 reflection tend to reduce more efficiently for higher annealing temperatures and longer annealing time spans.

The material quality of the annealed layers was examined by measuring the XRC-FWHM of the 0002 and 10-12 reflections. The results are shown in Fig. 3.12 and the related TDDs estimated by calculation are plotted in Fig. 3.13. The XRC-FWHM and related TDD of one of the as grown layers before annealing are also plotted as reference. However, it has to be considered that there always is a deviation present from run to run and also inside one run from wafer to wafer for nominally equal layers. Main reason for the the run-to-run deviation is the AlN nucleation being extremely sensitive to small changes in the reactor atmosphere as for example the conditioning of the reactor or the wear of the different reactor parts. The wafer-to-wafer deviation inside a single growth run is due to inhomogeneous heatcoupling of the different satellites to the susceptor. Another reason would be the condition of the different satellites in terms of for instance the amount of parasitic AlN at the edges of the satellites. The XRC-FWHM of the 0002 reflection do not show a clear trend. They are distributed within a range of 20″ which is in the mentioned range of the typical wafer-to-wafer reproducibility. For the XRC-FWHM of the 10-12 reflection, a clear trend is visible. For an annealing time of 1300 °C the XRC-FWHM is comparable to the as grown layer while for increasing annealing temperatures the XRC-FWHM decrease. Only for the temperature of 1550 °C a slight increase is observed. For longer annealing time spans, the XRC-FWHM of the 10-12 reflection also decreases for both temperatures

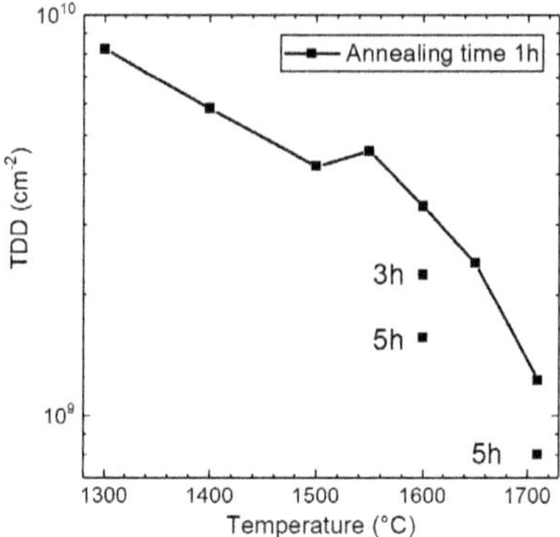

Figure 3.13: TDDs estimated by calculation from the XRC-FWHM of the 0002 and 10-12 reflections shown in Fig. 3.12 for 900 nm thick AlN layers annealed at different temperatures and for different time spans. The TDD is more efficiently reduced for higher annealing temperatures and longer annealing time spans.

of 1600 °C and 1710 °C. The TDDs follow the same trend as the XRC-FWHM of the 10-12 reflection. So it can be concluded that both a higher annealing temperature and a longer annealing time lead to a more efficient defect reduction. The only data point for which this is not valid is for the temperature of 1550 °C. This is due to the previously mentioned deviation for nominally equal layers, although they were prepared in one single growth run. The TDD reached after HTA is still dependent on the initial TDD. This was one of the outcomes in the previous section of this chapter. Considering this, the explanation for the value at 1550 °C is that the initial TDD was most likely higher than the value of the one as grown layer kept as reference. Thus, generally it can still be stated that for higher annealing temperatures and longer time spans the material quality improves. The lowest TDD reached was calculated to be $8 \times 10^8\,\mathrm{cm}^{-2}$ for the highest annealing temperature and longest time. The TDDs are a bit higher compared to the results of the thickness series in the previous section of this chapter. One reason for that could be that the annealing time in the furnace at HTM Reetz used for those experiments was always longer since heating up and cooling down took more time. The temperature transients of the two different furnaces are shown in chapter 2.1 for comparison. Additionally, the absolute temperature scale of the two different furnaces can differ since temperature in general is a physical parameter which is not easy to measure absolutely. Furthermore, the measured temperature strongly depends on the location of the respective thermocouple.

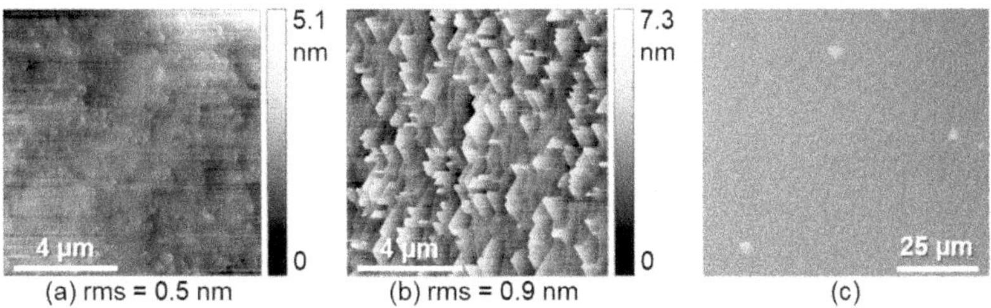

(a) rms = 0.5 nm (b) rms = 0.9 nm (c)

Figure 3.14: AFM images of 900 nm thick AlN layers after HTA at 1710 °C for **(a)** 1 h and **(b)** 5 h. **(c)** Light microscopy image with Nomarski contrast of the 900 nm thick AlN layer after HTA at 1710 °C for 5 h.

The surface morphology of the two layers annealed at 1710 °C for 1 h and 5 h with the lowest TDDs was checked by AFM (Fig. 3.14). The layer annealed for 5 h shows formation of step bunches similar to the layers in the previous section of this chapter. However, the step bunches are in a more faceted shape than compared to the annealed layers in Fig. 3.5. Due to their geometric arrangement the facets are ascribed to m-facets of the AlN. Considering the high temperature of 1710 °C applied for 5 h, the AlN adatoms have enough energy over a long enough period of time to form these facets for energy minimisation of the surface. The layer with an rms value of 0.9 nm (for an area of 10 µm x 10 µm) is still reasonably smooth. However, achieving an atomically smooth surface by regrowth in step-flow growth mode might be more difficult, since AlN m-facets tend to be stable once they formed. The layer annealed for only 1 h shows an atomically smooth surface with an rms value as low as 0.5 nm (for an area of 10 µm x 10 µm). Checking the macroscopic surface morphology by light microscopy imaging in Nomarski, small parts with a faceted shape were observed for the layer annealed at 1710 °C for 5 h. The bright contrast is ascribed to AlON formation at the AlN/sapphire interface, similarly observed for sputtered AlN/sapphire layers in reference [180]. Thus, an annealing temperature of 1710 °C applied for 5 h is supplying enough energy to the system that it starts to destabilise, which is not wanted. The layer annealed for only 1 h was checked by light microscope and did not contain any irregular contrast. Hence, for an annealing time of 1 h the AlN layer is chemically stable.

In the previous section of this chapter, a change of the strain due to HTA was derived from curvature measurements (Fig. 3.11). To quantitatively analyse this change of the strain at room temperature in dependency of annealing temperature and time, high-resolution XRD 2θ scans of the symmetric reflections 0004 and 0006 were measured

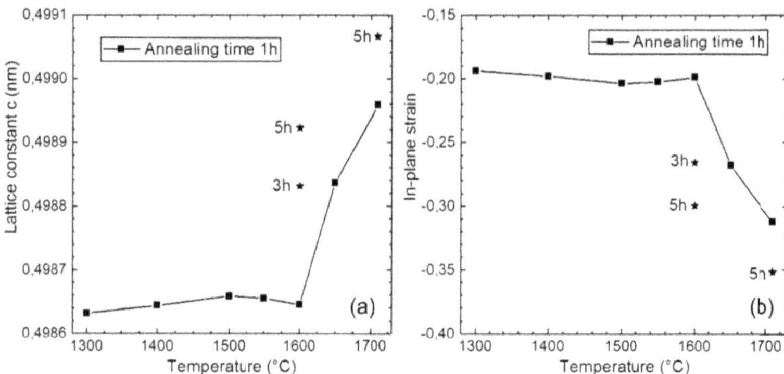

Figure 3.15: (a) Lattice constant c and **(b)** calculated in-plane strain at room temperature of 900 nm thick AlN layers after HTA at different temperatures and time spans assuming biaxial strain.

by Eberhard Richter at FBH. The out-of-plane c lattice constant at room temperature can be derived from the angular peak position of such scans as described in chapter 2.2. The determined c lattice constants for the layers of the temperature and time series are plotted in Fig. 3.15a. The c lattice constant at room temperature increases with higher annealing temperatures and also longer annealing time spans. This effect becomes very distinct for temperatures above 1600 °C and also for longer annealing time spans at 1600 °C and 1710 °C. Assuming biaxial strain, an increase in the c lattice constant translates into a reduction of the in-plane a lattice constant. This leads to a change of the in-plane strain to more compressive values which by definition have a negative sign. The in-plane strain at room temperature was calculated from the c lattice constant as described in chapter 2.2 and is plotted in Fig. 3.15b. The absolute value of the in-plane strain follows the trend of the c lattice constant. Similar results were observed by different research groups for sputtered but also MOVPE grown layers with a post growth HTA step [116,128]. This phenomenon was more closely investigated by Wang et al. and can be explained with a combination of the following two aspects:

1. The lattice mismatch between AlN and sapphire is temperature dependent due to their different thermal expansion coefficients. If enough energy is applied to the system via high annealing temperature for a certain amount of time, a reordering of the AlN/sapphire interface takes place resulting in a regular arrangement of misfit dislocations in order to compensate for the lattice mismatch present

at the high temperature, so that an approximately strain-free situation at the high temperature is achieved [128]. During cool-down, the arrangement of the AlN/sapphire interface will be conserved and compressive strain will build-up due to the thermal mismatch between AlN and sapphire. The compressive strain at room temperature becomes higher for higher annealing temperatures, since the temperature difference between the strain-free situation at high temperature and room temperature increases. Hence, more compressive strain due to the thermal mismatch will build-up.

2. The initial strain in the as grown AlN layers is tensile at growth temperature as could be seen in the previous sections of this chapter. The tensile strain is ascribed to low-angle grain boundaries forming at coalescence of the nucleation islands to compensate their tilt and twist [184]. These low-angle grain boundaries consist of networks of TDs. Hence, in this case the reduction of the TDD during HTA shifts the overall strain towards compressive values at the annealing temperature and thus, also at room temperature.

The model with these two effects is in good agreement with the data in Fig. 3.15: For higher annealing temperatures both effects play a role, while for longer annealing time spans the second effect of further reducing the TDD becomes more relevant. The change of the strain is a general drawback considering the HTA method for preparing AlN templates, since it makes relaxation of subsequently grown AlGaN more likely [185]. However, there has been already promising work how to handle the compressive strain induced by HTA to suppress relaxation in AlGaN layers grown on HTA AlN templates [123, 186]. On the other hand, if compressive strain is wanted, this can be achieved by applying HTA. This could be useful for growing thick AlN layers without cracking.

Summary In this chapter, applying HTA to MOVPE grown AlN templates in face-to-face configuration was investigated. The material quality was improved by annealing with the lowest TDD reached being $6 \times 10^8\,\mathrm{cm}^{-2}$ as determined by XRD. However, if the tensile strain incorporated during AlN growth is too high, the layer will crack during annealing. This obstacle can be overcome by modulation of the growth conditions to induce roughening and subsequent smoothing of the surface. Layers of different thickness prepared by this method did not show any cracks after annealing. The optimal layer thickness for subsequent annealing was found to be 900 nm. On one hand, the layer should be thick enough so that TDD reduction by dislocation annihilation with increasing layer thickness can take place, since the achieved TDD after annealing is still dependent on the initial TDD. On the other hand, if the layer gets too thick, the build-up of tensile strain can not be sufficiently suppressed so that layer cracking will occur. An advanced temperature and time series of the 900 nm AlN layer showed that the TDD can be more efficiently reduced by both increasing the annealing temperature and annealing time. The reduction of the TDD was approved by TEM investigations which further revealed the formation of microscopic AlON "bubbles" inside the AlN layer. This was related to oxygen diffusion from the sapphire substrate into the AlN layer leading to an increase in the oxygen concentration. Although high oxygen impurity concentrations can lead to absorption in the UV range, no impairment of the UV transmittance of the annealed AlN was found. For the highest annealing temperature of 1710 °C and longest annealing time of 5 h macroscopic formation of AlON was observed. Hence, the whole layer system starts to become chemically unstable. In general, the strain in the AlN layers is shifted towards compressive values after annealing. This effect becomes more dominant for increasing the annealing temperature and/or annealing time, which is attributed to two effects: The AlN/sapphire interface relaxes at the high temperature according to the temperature-dependent lattice mismatch. This results in the build-up of compressive strain during cool-down due to the thermal mismatch between AlN and sapphire. Additionally, reducing the TDD will further reduce the amount of tensile strain inside the layer which also points towards more compressive strain for the whole layer at annealing temperature and also at room temperature. The compressive strain causes an increase of the wafer bow after annealing. Furthermore, regrowth on HTA layers leads to build-up of more compressive strain rather than the usually observed tensile strain, which can be observed in the in-situ curvature measurement.

In conclusion, the method of HTA annealing of MOVPE grown AlN templates is a promising technique for preparing UV LED base layers. The only main drawback is the change of the strain to compressive values, because this will challenge the strain management of subsequently grown AlGaN heterostructures. However, from a general point of view it is interesting to note, that depending on the annealing temperature and time the TDD and also the strain in the AlN layer can be controlled. This for instance enables thick and crack-free AlN growth which might be desired for certain applications. A more general limitation of the HTA of AlN on planar sapphire is that it only deals with the improvement of the material quality and does not have any potential for increasing the light extraction efficiency of LED heterostructures grown on top.

However, increasing the light extraction efficiency is very important along the way to high performance UV LEDs. Another technique, which is commonly used for GaN-based LEDs to improve both the material quality and the light extraction efficiency, is the growth on NPSS. The AlN growth on such substrates will be investigated in detail in the next chapter of this work.

CHAPTER 4

Nanopatterned sapphire substrates

Epitaxial lateral overgrowth (ELO) of patterned sapphire substrates is a common technique in III-nitride technology for fabrication of high quality pseudo substrates. However, while for GaN the technique is well established and used in commercial mass production for GaN-based blue LEDs, in case of AlN additional problems arise (details described in chapter 1.3): The high sticking coefficient of Al leads to AlN growth on all sapphire facets resulting in overall polycrystalline growth. Hence, for the desired final c-plane AlN, the sapphire pattern firstly needs to contain enough c-plane. Secondly, the c-plane growth has to be favoured by appropriate growth conditions so that after coalescence only the c-plane AlN remains without any disturbances due to parasitic non c-planar growth on the sapphire side facets. The first realization of AlN ELO succeeded by using a one-dimensional stripe pattern in the micrometre range [66]. However, recently there has been increasing research interest in using 2-dimensional patterns in the nanometre range [94, 97, 98]. The motivation for that research are the following potential advantages over the commonly used stripe pattern:

- Since for AlGaN-based UV LEDs the generated light is coupled out through the sapphire substrate, the AlN/sapphire interface is crucial regarding the light extraction efficiency (LEE). Usually, a huge portion of the light is lost at the interface due to total internal reflection. This can be suppressed by implementing a patterned interface. While only a slight increase of the LEE is expected for the 1-dimensional stripe pattern in the micrometre range, a 2-dimensional nanopattern is believed to have the potential of a distinct increase, because it could work similar to a photonic crystal structure. However, reports in literature of a direct proof of such an effect are very limited [99, 187].

- Smaller pattern dimensions lead to a reduced coalescence thickness which is beneficial in terms of wafer curvature and production costs.
- Due to the small dimensions of the pattern, TDs at growth start have enough energy at growth temperature to bend towards free surfaces and end there. This leads to an overall more efficient dislocation reduction.
- The use of 2-dimensional patterns instead of 1-dimensional patterns results in improved isotropy in terms of light extraction and strain relaxation. 2-dimensional patterns in the micrometre range are tough to coalesce due to formation of stable AlN m-facets inhibiting coalescence. For 2-dimensional patterns in the nanometre range the layers eventually coalesce before the stable AlN m-facets fully form.

While these potential advantages lay a proper foundation for the research work, they will be discussed and compared with new findings along the results of the following investigations.

In the first part of this chapter, wave optical simulations are presented to quantify the potential increase of the light extraction of UVC LEDs with an implemented nanopatterned AlN/sapphire interface. The second and third part of this chapter deal with the challenges along the way to the successful AlN overgrowth of two different nanopatterns (nanopillars and nanoholes). Main criterion for successful overgrowth is the suitability of the final layer as a UV LED template in terms of material quality and surface morphology.

4.1 Light extraction simulations of nanopatterned interface

As introduced, it is widely believed in the LED community that implementing a nanopatterned AlN/sapphire interface increases the LEE of the final UV LED. However, direct proof of this hypothesis in literature is very limited, since the LEE can not be measured directly. The increase in the output power of UV LEDs on patterned substrates can always be due to both the improved material quality and the enhanced LEE making conclusion about the LEE in isolation difficult. To get access to the behaviour of the light extraction, wave optical simulations were conducted in collaboration with P. Manley at the Helmholtz Zentrum Berlin. For the wave optical simulations, scattering

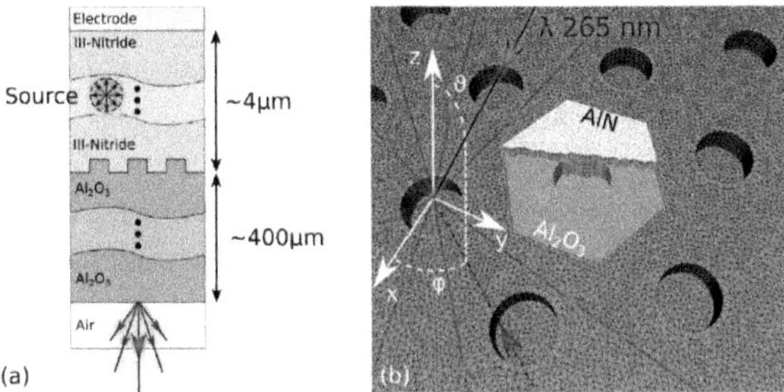

Figure 4.1: (a) Sketch of the simplified layer stack of the UV LED structure used for the simulation. The light source with wavelength of 265 nm was placed inside the III-nitride layer at the position of the active region. **(b)** Mesh grid of the nanopatterned AlN/sapphire interface in the simulation setup. The nanopatterns are of hexagonal symmetry. The unit cell is highlighted and only partly filled with AlN for better visibility. The 265 nm light propagates from the AlN to the sapphire. The incident angle is determined by its polar angle θ and azimuthal angle ϕ. From reference [162] (second authorship) with permission of The Optical Society (USA) © 2020.

matrices of the material interfaces were determined and fed into a system of equations to describe the light propagation inside the layer stack of the UV LED structure. The coupling coefficients of the scattering matrices were calculated by the finite element method. Details of the simulation procedure are reported in reference [162]. The simulation setup containing a sketch of the simplified UV LED structure and the mesh grid of the nanopatterned AlN/sapphire interface are shown in Fig. 4.1. 265 nm was chosen for the wavelength of the UV LED and the source of the light was placed inside the simplified layer stack at the position of the active region of the LED. The main assumptions made for the simulation are the following:

- Since the refractive indices do not change much inside of the actual AlGaN layer stack representing the UV LED, all AlGaN layers are implemented as one layer with the refractive index of AlN. Several layers and the resulting interfaces would dramatically increase the computational time.

- Coherence effects of the light inside the layer structure are neglected since the two leftover layers of the simplified structure are both thick enough that the light will loose coherence for multiple passages through the structure.

- The AlN is assumed to fully cover the nanopatterned sapphire so that no air void inclusions are present. Air voids are often observed for AlN overgrowth of patterned sapphire. However, they are neglected here to keep the computational time at a reasonable level.

These assumptions simplify the simulation so that the computational time needed becomes feasible. The absolute values of the simulation results might differ from reality due to these assumptions. However, the general trends should not be altered, so that drawing conclusions is still reasonable.

As a first step, the transmittance of the light through the AlN/sapphire interface is calculated for only a single light pass inside the UV LED structure. This corresponds to an LED with a non-transparent p-side, which is up to now the usual case for UV LEDs emitting at 265 nm [62]. GaN is still typically used as the final contact layer to ease the problem of p-type conductivity. P-type conductivity is a general problem in III-nitrides and becomes even more difficult for AlGaN. However, the GaN contact layer leads to absorption of light with a wavelength below 360 nm. So one half of the generated light will be absorbed and also the reflected light at the AlN/sapphire interface passing the structure more than once will be absorbed. The AlN/sapphire interface is implemented in one case as a planar interface as reference and in the other cases consisting of different nanopatterns (Fig. 4.2a-c). The different nanopatterns are all of hexagonal symmetry and consist of either nanoholes, nanopillars or nanofrustums. Their dimensions are varied with the pitch P and respective variation of the diameter $D = P/2$ and height $H = P/4$. The single pass transmittance from AlN into sapphire for the different nanopatterned AlN/sapphire interfaces and also the planar interface as reference is plotted against the polar angle θ of the incident light in Fig. 4.2d-f. The azimuthal angular dependence has been numerically integrated out in order to focus on the polar angular dependence. Due to absorption of one half of the light in the non-transparent p-side of the LED, the maximum value for the transmittance is 0.5. The polarisation of the light is implemented as isotropic, i.e. 50 % transversal electric (TE) and 50 % transversal magnetic (TM). In reality, the light polarisation of a UV LED strongly depends on the characteristics of the MQW. For an emission at 265 nm of pseudomorphically grown MQW the light should be almost fully TE polarised [188]. However, for reason of generalisation regarding UV LEDs of different wavelengths the polarisation is kept isotropic. The difference in transmittance for pure TE and pure TM polarised light will be discussed later in this section. The first thing to note is that above the critical angle of 51° there is no transmittance for the planar interface due to total internal reflection. For the nanopatterned interfaces this transition flattens so that transmittance is gained above the critical angle but lost below it. This flattening effect becomes more pronounced for higher pitches of the nanopatterns. Nanoholes, nanopillars and nanofrustums behave almost identical and thus, the shape of the nanopattern itself

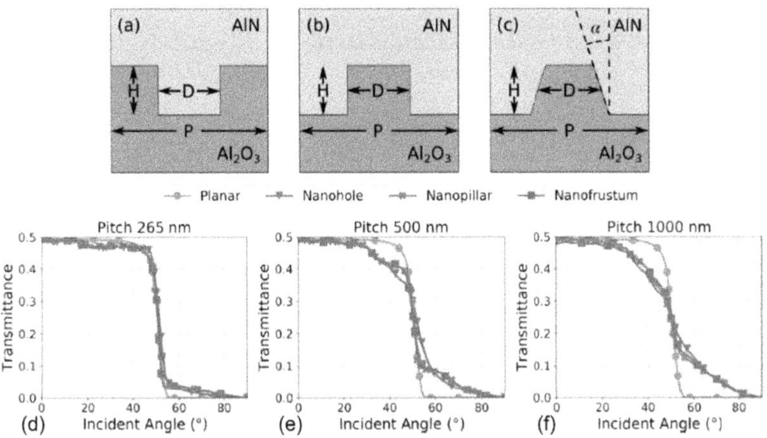

Figure 4.2: Cross-sectional sketches of the different nanopatterned interfaces consisting of **(a)** nanoholes **(b)** nanopillars and **(c)** nanofrustums with dimensions pitch P, diameter $D = P/2$ and height $H = P/4$. Incident polar angle θ of the generated light (isotropic light polarisation) plotted against the single pass transmittance for different AlN/sapphire interfaces including a planar interface as reference and the nanopatterned interfaces of **(a)**-**(c)** with pitches of **(d)** 265 nm **(e)** 500 nm and **(f)** 1000 nm. Since the single pass transmittance case corresponds to an LED with a non-transparent p-side, a maximum transmittance of 0.5 is obtained. The azimuthal angular dependence has been numerically integrated out to focus on the polar angular dependence. From reference [162] (second authorship) with permission of The Optical Society (USA) © 2020.

does not have a big influence. The integrated transmittance for the curves in Fig. 4.2d-f were calculated and are shown in Tab. 4.1a. The integrated transmittance from AlN into sapphire shows an increase for all nanopatterned interfaces in comparison to the planar interface. However, the increase is relatively small. The highest value translates into an increase of 9 % and is reached for the interface consisting of nanoholes with a pitch of 1000 nm. In Tab. 4.1b+c the integrated transmittances are shown for pure TE and pure TM polarised light to get an idea of the influence arising from differently polarised light. For pure TE polarised light the overall transmittance increases: Since emission occurs more parallel to the z-direction, less light is lost due to total internal reflection. The difference between the planar interface and the nanopatterned interface becomes smaller. The highest value translates into an increase of 5 % and is reached for the interface consisting of nanoholes with pitch of 1000 nm. For pure TM polarized light the opposite effect is observed: More light is emitted perpendicular to the z-direction and thus, more light is lost due to total internal reflection and the overall integrated

Table 4.1: Integrated single pass transmittance from AlN into sapphire for the planar interface and the interfaces consisting of the different nanopatterns for **(a)** isotropic light polarisation, **(b)** purely TE and **(c)** purely TM polarised light. From reference [162] (second authorship) with permission of The Optical Society (USA) © 2020.

(a) Isotropic light polarisation

Pitch (nm)	Planar	Nanoholes	Nanopillars	Nanofrustums
265	17.6 %	18.6 %	18.3 %	18.3 %
500	17.6 %	18.6 %	18.7 %	18.6 %
1000	17.6 %	19.2 %	18.9 %	19.0 %

(b) Purely TE polarised light

Pitch (nm)	Planar	Nanoholes	Nanopillars	Nanofrustums
265	22.1 %	23.0 %	22.6 %	22.5 %
500	22.1 %	22.8 %	22.8 %	22.7 %
1000	22.1 %	23.1 %	22.8 %	22.9 %

(c) Purely TM polarised light

Pitch (nm)	Planar	Nanoholes	Nanopillars	Nanofrustums
265	8.4 %	9.7 %	9.6 %	9.6 %
500	8.4 %	10.3 %	10.5 %	10.4 %
1000	8.4 %	11.4 %	11.0 %	11.2 %

transmittance reduces. In this case, the difference between the planar interface and the nanopatterned interfaces becomes bigger. The highest value translates into an increase of 36 % and is reached for the interface consisting of nanoholes with pitch of 1000 nm.

In the following simulations, the p-side is no longer assumed to be fully absorbing. Instead, a mirror with an effective reflectivity is introduced at the p-side to enable multiple light passages through the structure. Since there was no significant difference for the nanoholes, nanopillars and nanofrustums in case of the single pass transmittance, the multiple pass transmittance is only simulated for the nanoholes without loss of generality. The pitch was varied analogous to the previous simulations for the single pass transmittance. Furthermore, the reflectivity of the mirror at the p-side was varied from 0 up to 1. The schematic layer system and the resulting curves are depicted in Fig. 4.3. The reflectivity of the mirror set to 0 corresponds to the single pass transmittance and hence, the difference between planar interface and nanopatterned interface is only very small. Increasing the reflectivity leads to an increase of the transmittance through both interfaces. However, while the transmittance through the planar interface increases

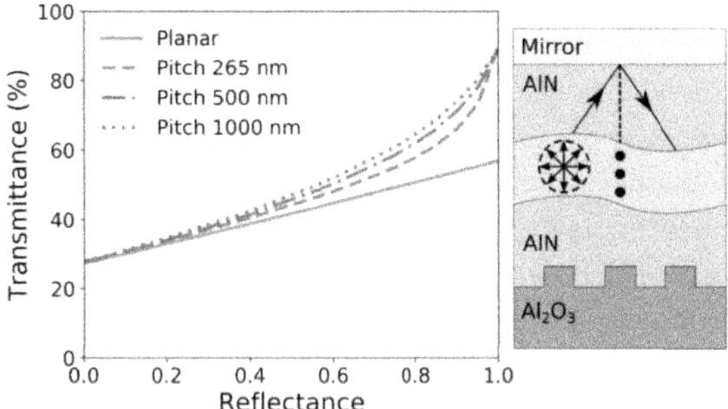

Figure 4.3: Transmittance of light with 265 nm wavelength from AlN into sapphire as a function of the reflectivity of the mirror placed at the p-side to induce an effective reflectivtiy of the p-side. The schematic structure of the layer system is shown in the small sketch to the right. From reference [162] (second authorship) with permission of The Optical Society (USA) © 2020.

only linearly, the transmittance through the nanopatterned interface increases more strongly and is of nonlinear shape. For the higher pitch of the nanopattern the increase happens faster, but all curves are converging to the same value for a reflectivity of 1. In the theoretical limit the transmittance would be 1 since all kinds of absorption in the layer structure are neglected. However, in the simulation the computational time dramatically rises due to the increasing number of light passages. Hence, the simulation was stopped for a reflectivity close to 1 with related transmittance of 0.89 as can be seen in Fig. 4.3. In contrast, the maximum value reached for the planar interface is still limited to 0.57, since total internal reflection fully traps a big portion of the generated light inside the layer system. This is not the case for the nanopatterned interface due to a non-zero transmittance even above the critical angle of total reflection with respect to the planar interface observed for the single pass transmittance (Fig. 4.2). Thus, increasing the number of light passages inside the layer system, transmittance of the light above the critical angle will eventually happen. This effect happens faster for the nanopatterns with higher pitches, since the probability for transmittance above the critical angle is higher. Coming back to the experimental reality, a transparent AlGaN p-side for a UV LED without the fully absorbing p-GaN contact layer can in principle be achieved by sacrificing electrical efficiency. Combined with a reflective p-contact material such as Ni/Mg or Rh electrodes a value of 0.8 for the effective reflectivity of the p-side can be implemented in a realistic scenario [189, 190]. For an effective reflectivity of 0.8 the simulation shows an increase of the transmittance from the AlN into sapphire of 0.59 for the nanopatterned interface consisting of nanoholes with pitch of 1000 nm in contrast to a transmittance of only 0.33 for the planar interface. This translates to a relative increase of 77 % for the transmittance through the nanopatterned AlN/sapphire interface in comparison to the planar one. However, it has to be considered that this is only the transmittance into the sapphire. For the complete light extraction the sapphire/air interface at the sapphire backside also plays an important role and might change the simulated increase ascribed to the nanopatterned interface. The sapphire backside usually is not polished and therefore rough to increase the light extraction by enhanced scattering. Another approach for a more controlled enhancement of the light extraction into air at the sapphire/air interface is the patterning of the sapphire backside [191].

The simulation results show that there is a potential increase in the light extraction by implementing a nanopattern at the AlN/sapphire interface. While this effect becomes almost negligible for an LED with an absorbing p-side, it should be very beneficial for an LED with transparent p-side and reflecting contact. The route to high efficiency UV LEDs inevitably leads to the development of such structures, since loosing 50 % due to absorption in the p-side is not feasible. Hence, there is a big motivation for overgrowing nanopatterned sapphire by AlN not only to improve the material quality, but also to improve the light extraction efficiency. The following two sections will deal with the development growth techniques for overgrowing such nanopatterns.

4.2 Growth on sapphire nanopillars

The first nanopattern used for overgrowth consists of hexagonally aligned nanopillars. The advantage of this pattern lies in the independent nucleation of the AlN on each pillar similar to the stripe pattern so that the build-up of tensile strain and resulting cracking of the layer should not be an issue. The starting point of this work is the growth technique for coalescence of AlN grown on sapphire nanopillars, which was developed by Hagedorn et al. at FBH in the same growth reactor used for this work [180]. However, in the cited work the coalesced layer showed high step bunches (> 10 nm) on the surface. This can be detrimental for further AlGaN epitaxy causing inhomogeneity in composition [192]. Formation of step bunches depends on the offcut of the sapphire substrate surface. Hence, the influence of the offcut variation on the AlN overgrowth and further AlGaN epitaxy will be investigated. The offcut used by Hagedorn et al. was 0.2° towards the m direction which is the standard sapphire offcut for industrial mass

Figure 4.4: SEM images of the fabricated hexagonally aligned nanopillars with pitch of 1000 nm **(a)** in bird's-eye view and **(b)** in cross-sectional view. From reference [134] (first authorship) with permission of Elsevier © 2020.

production. A reduced offcut of 0.1° towards the m direction is chosen for comparison, since a reduced offcut should shift the growth mode in the direction of step-flow growth and thus, lead to a smoother surface [175, 176]. Additionally, a reduced offcut leads to smaller height differences of adjacent AlN columns on the nanopillars. This results in smaller steps at their coalescence [193].

The sapphire nanopillars were prepared by Pierre-Marie Coulon at the University of Bath applying Displacement Talbot Lithography and subsequent dry etching to the sapphire substrates. Details of the fabrication process are described in chapter 2.1. SEM images of the final nanopatterned substrates are shown in Fig. 4.4 in bird's-eye view and cross-sectional view. A homogeneous nanopattern consisting of hexagonally aligned nanopillars with pitch of 1000 nm was achieved over the full 2-inch wafer. Nominally equal nanopatterns were fabricated on sapphire substrates with offcuts of 0.2° and 0.1° towards the m direction. The overgrowth mainly consists of three steps:

1. Standard nucleation at low temperature of 980 °C.
2. High temperature growth to promote coalescence at 1380 °C.
3. Medium temperature growth for smoothing by step-flow growth at 1180 °C.

The in-situ measurement of the 405 nm reflectance, the process temperature and the curvature is shown in Fig. 4.5 for the overgrowth of both the 0.2° and 0.1° offcut sample. The samples were overgrown in different growth runs following the same method. After standard thermal cleaning and standard nucleation, the growth temperature was increased to the maximum temperature of the growth reactor of 1380 °C. The V/III ratio was set to 30. Under such conditions the diffusion length of the Al-adatoms on the growth surface is very high and hence, lateral growth to coalesce the layer is favoured.

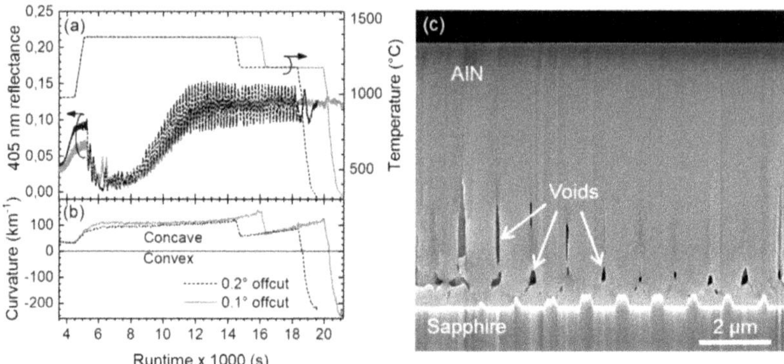

Figure 4.5: (a) 405 nm reflectance, process temperature and **(b)** curvature against run time for the three-step growth process to overgrow the sapphire nanopillars with offcut of 0.2° (dashed-black) and 0.1° (solid-red). **(c)** Cross-sectional SEM image of the overgrown sapphire nanopillars exemplarily shown for the 0.1° offcut sample. Coalescence of the layer is achieved after approx. 4 µm.

The reflectance shows Fabry-Pérot oscillations due to increasing layer thickness. The average reflectance at first decreases after nucleation, but then starts to increase until saturation occurs at a value of about 0.12 indicating coalescence of the layer. Starting from the nucleation islands on the nanopillars, the AlN initially grows in the shape of a sphere on each nanopillar with only a small amount of remaining *c*-facet on the top [97]. This causes the decrease of the average reflectance in the first stage of growth after nucleation. The *c*-facet then builds up leading to the increase of the average reflectance until coalescence is reached. The thickness of coalescence is approximately 4 µm for both samples with different offcut. However, the point in time to switch to the medium temperature was set manually to ensure coalescence is reached by waiting for the average reflectance to saturate. For that reason, the high temperature growth step was applied approximately 10 min longer for the 0.1° offcut sample, although coalescence seems to happen similar for both samples. The following growth at medium temperature of 1180 °C at the same V/III ratio of 30 is introduced to induce step-flow growth with the aim to end up with an atomically smooth surface. The total layer thickness is approximately 6.6 µm for the 0.2° offcut sample and 6.9 µm for the 0.1° offcut sample.

The curvature (Fig. 4.5b) increases after nucleation following the temperature ramp up to the high temperature of 1380 °C. It slightly increases during the high temperature growth but then stays relatively constant until coalescence is completed. The increase in the beginning of the high temperature growth is most likely due to AlN growth on the planar sapphire between the nanopillars at growth start. As soon as the spheres on top of the nanopillars build up, less material reaches the planar parts between the nanopillars. The non-continuous AlN on the sapphire nanopillars grows strain-free until the closed layer forms at coalescence. From there on, the growth is tensile again leading to a linear increase in the curvature. This linear increase is only interrupted by the abrupt decrease due to the temperature ramp down to the medium temperature of 1180 °C. While cooling down the curvature strongly decreases switching to convex values of $-210\,\mathrm{km}^{-1}$ for the 0.2° offcut sample and $-225\,\mathrm{km}^{-1}$ for the 0.1° offcut sample. As explained in the previous chapter, this effect is related to the thermal mismatch between AlN and sapphire. The curvature qualitatively follows the same trends for the two samples with different offcut. Quantitatively, the 0.1° offcut sample has a more convex shape after cool down due to the larger layer thickness. The absolute curvature values are relatively high for both samples. This is related to the high AlN layer thickness and the resulting stronger impact of the thermal mismatch. In Fig. 4.5c, a cross-sectional SEM image is exemplarily shown for the 0.1° offcut sample to analyse the structural morphology. Voids can be observed which penetrate through the layer until full coalescence of the layer is achieved. Despite the small pattern pitch of 1000 nm, coalescence is completed relatively late after growth of around 4 µm thick AlN. After fast lateral growth shortly after nucleation, full coalescence happens only after several micrometres of side-by-side growth. This leads to the conclusion that the final step of coalescence is inhibited and becomes unlikely.

The surface morphology of both samples with different offcut before and after the medium temperature growth step for smoothing the surface was investigated by AFM (Fig. 4.6). Before smoothing, there are step bunches visible for both samples (Fig. 4.6a,b) resulting in a relatively rough surface with rms values (for an area of 10 µm x 10 µm) of 3.6 nm for the 0.2° and 2.9 nm for the 0.1° offcut sample. Step bunches form, if the Al-adatom diffusion length exceeds the terrace width, which is determined by the offcut [175]. Thus, the smaller terrace width caused by the higher offcut of 0.2° induces a more efficient formation of step bunches. The result is higher step bunches and a higher rms value in comparison to the 0.1° offcut sample. After the medium temperature growth, the 0.2° offcut sample still shows step bunches and the rms value (3.7 nm for an area of 10 µm x 10 µm) stays unchanged. From chapter 3.2 it is known that the applied growth conditions with temperature of 1180 °C and V/III ratio of 30 and an offcut of 0.2° induce step-flow growth. However, even after growth of about

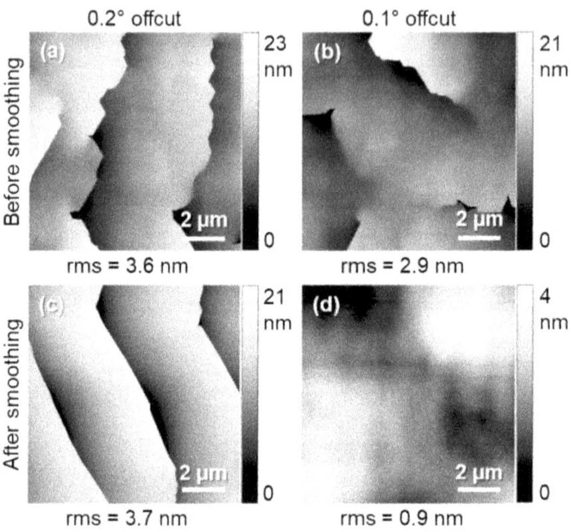

Figure 4.6: AFM images of the AlN surface after overgrowing the sapphire nanopillars before and after smoothing by the medium temperature growth step at 1180 °C and V/III ratio of 30 for **(a)**, **(c)** the 0.2° offcut sample and **(b)**, **(d)** the 0.1° offcut sample. The rms value is included for better comparison of the overall roughness. While for the 0.2° offcut sample, the step bunches remain even after smoothing, for the 0.1° offcut sample, an atomically smooth surface is achieved. From reference [134] (first authorship) with permission of Elsevier © 2020.

2 µm AlN in step-flow growth the step bunches could not be smoothed. They still have a height of about 10 nm leading to the conclusion that such high step bunches can not be smoothed by simply applying growth in step-flow growth mode. For the 0.1° offcut sample, an atomically smooth surface (rms = 0.9 nm for an area of 10 µm x 10 µm) with bilayer steps is achieved after the medium temperature growth step (Fig. 4.6c,d). Thus, the height of the step bunches of the 0.1° offcut sample was not too high for being smoothed by the 2 µm AlN in step-flow growth.

For defect analysis, the standard technique of measuring the XRC-FWHM of the 0002 and 10-12 reflections was carried out. For the 0.2° offcut sample values of 300″ and 810″ and for the 0.1° offcut sample values of 510″ and 1160″ were determined for the 0002 and 10-12 reflections, respectively. The estimated TDDs were calculated to be $7 \times 10^{9}\,\mathrm{cm}^{-2}$ for the 0.2° and $1 \times 10^{10}\,\mathrm{cm}^{-2}$ for the 0.1° offcut sample. These values are surprisingly high for AlN on sapphire of thickness larger than 6 µm. The

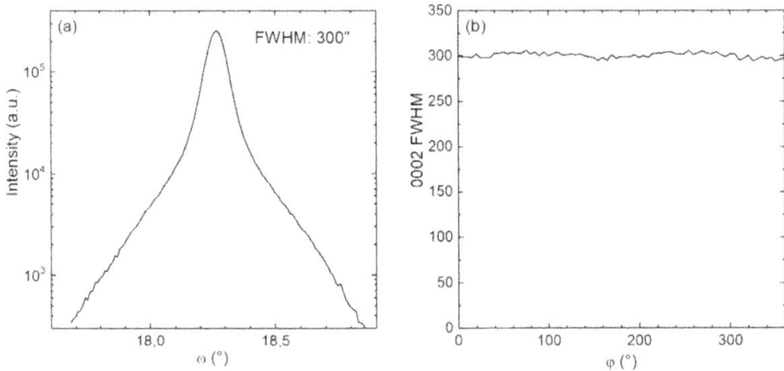

Figure 4.7: (a) X-ray ω rocking curve of the 0002 reflection for the 0.2° offcut sample after AlN overgrowth and **(b)** XRC-FWHM of the 0002 reflection against rotational angle ϕ. The broadening of the XRC-FWHM due to wingtilt is present in all directions and therefore cannot be mitigated by adapting the measurement geometry.

calculation of the TDD from XRC-FWHM will be only reliable, if no other significant broadening of the XRC-FWHM is present. Broadening from other sources than the TDs will falsify the TDD estimated from the XRC-FWHM. In literature there are several reports about the occurrence of wing tilt in case of epitaxial lateral overgrowth of non-continuous patterns in the laterally overgrown parts of the layer [194,195]. Wing tilt leads to broadening of the XRC-FWHM and is therefore most likely the reason for the relatively high values for the TDDs calculated from the XRC-FWHM. Analogous effects are known for overgrown sapphire stripe patterns. However, in case of the 1-dimensional stripe pattern, the XRD-FWHM are only broadened in one dimension perpendicular to the strips. By adapting the measurement geometry and measuring parallel to the stripes, the broadening related to wing tilt can be mitigated [97]. For the 2-dimensional, hexagonally aligned nanopillar pattern, broadening is present in all directions. This was exemplarily checked for the 0.2° offcut sample by measuring the XRC-FWHM of the 0002 reflection in dependence of the rotational angle ϕ (Fig. 4.7). The broadened rocking curve is depicted in Fig. 4.7a while in Fig. 4.7b it is proven that the XRC-FWHM of the 0002 reflection does not significantly change by simply rotating the sample. Thus, it is not possible to assess the TDD by the standard technique of estimation from the XRC-FWHM.

Since the determination of the TDD is very crucial to evaluate the AlN template quality, the alternative methods of electron channelling contrast imaging (ECCI) and

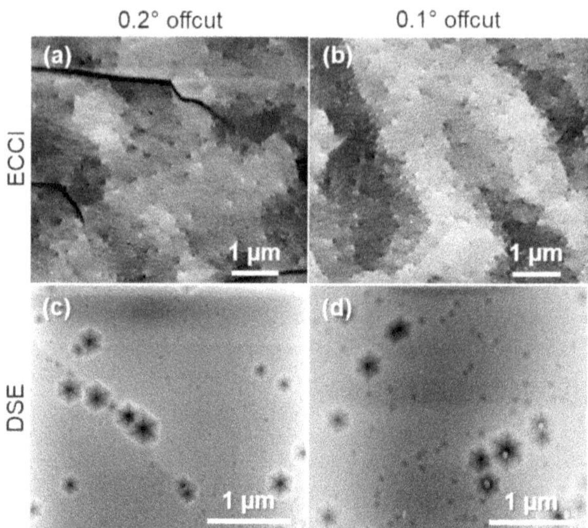

Figure 4.8: ECCI images of the surface for the **(a)** 0.2° and **(b)** 0.1° offcut sample and SEM images of the surface after DSE for the **(c)** 0.2° and **(d)** 0.1° offcut sample. The TDDs determined by these more advanced methods suggest lower TDDs than estimated from the broadened XRC-FWHM. From reference [134] (first authorship) with permission from Elsevier © 2020.

defect-selective etching (DSE) were conducted for both samples (Fig. 4.8) in collaboration with other research groups specialised in these methods (details described in chapter 2.2). ECCI was performed by Gunnar Kusch, Aeshah Alasmari and Carol-Trager Cowan at the University of Strathclyde. The achieved SEM images can be found in Fig. 4.8a,b. The TDs are visible as point-like contrast caused by diffraction of the electron beam under certain geometric conditions [156,157]. The density of the point-like contrasts directly corresponds to the TDD and is counted to be $8 \times 10^8\,\mathrm{cm}^{-2}$ for the 0.2° and $1 \times 10^9\,\mathrm{cm}^{-2}$ for the 0.1° offcut sample. Additionally, DSE was carried out by Nicole-Yvonne Suss, Lucinda Matiwe and Carsten Hartmann at IKZ. SEM images of the etched surfaces are presented in Fig. 4.8c,d. The etch pit density directly corresponds to the TDD and is counted to be $9 \times 10^8\,\mathrm{cm}^{-2}$ for the 0.2° and $1 \times 10^9\,\mathrm{cm}^{-2}$ for the 0.1° offcut sample. The size of the etch pits is ascribed to the different types of dislocations: The small etch pits are related to pure edge type dislocations while the bigger etch pits are related to mixed type and pure screw type dislocations [159,160]. In Tab. 4.2 the resulting TDDs for the different methods are shown for comparison.

Table 4.2: TDDs determined by estimation from the XRC-FWHM, ECCI and DSE for the 0.2° and 0.1° offcut sample. While from the XRC-FWHM the TDDs are overestimated due to wingtilt, the values obtained by ECCI and DSE are more reliable.

Characterisation method	TDD 0.2° offcut sample	TDD 0.1° offcut sample
XRD	$7 \times 10^9\,\mathrm{cm}^{-2}$	$1 \times 10^{10}\,\mathrm{cm}^{-2}$
ECCI	$8 \times 10^8\,\mathrm{cm}^{-2}$	$1.5 \times 10^9\,\mathrm{cm}^{-2}$
DSE	$9 \times 10^8\,\mathrm{cm}^{-2}$	$1 \times 10^9\,\mathrm{cm}^{-2}$

The small deviation in the TDD for the 0.2° offcut sample between ECCI and DSE can be explained by the overlap of contrast from the TDs and the edges of the step bunches for ECCI (Fig. 4.8a). Thus, TDs at the step edges are not counted which leads to a slight underestimation of the overall TDD. In the DSE image it is observed, that TDs tend to line up along the edges of the step bunches (Fig. 4.8c). Since TDs are attracted from free surfaces by the related image forces, they bend towards the sidewalls of the step bunches during growth resulting in the observed lining up along the step edges [57]. This also explains the slightly lower TDD for the 0.2° offcut sample compared with the 0.1° offcut sample, because the lining up makes additional dislocation annihilation more likely due to a reduction of the average distance between the TDs. For the uniformly distributed dislocations of the 0.1° offcut sample the TDDs determined by ECCI and DSE are in good agreement. Overall, it can be concluded that the TDDs determined by DSE and ECCI are more reliable than the TDDs estimated from the XRC-FWHM. Consequently, much lower TDDs of around $1 \times 10^9\,\mathrm{cm}^{-2}$ are obtained for the sapphire nanopillars overgrown by AlN.

To evaluate the influence of the offcut variation on subsequent AlGaN epitaxy, both samples were overgrown with AlGaN MQW emitting at 265 nm at room temperature. The AlGaN heterostructure was grown in a close-coupled showerhead reactor at the Technical University of Berlin by Norman Susilo. The applied growth recipe is identical to the one described in reference [107]. After growth, the luminescence behaviour of the MQW was investigated by spatially resolved low temperature CL at 80 K with acceleration voltage of 5 keV for the electron beam. The measurement was carried out by Carsten Netzel at FBH and is shown in Fig. 4.9. The main luminescence peak for both samples occurs at a wavelength of 255 nm and is attributed to the luminescence of the MQW. At room temperature the peak wavelength lies at 265 nm, but due to the temperature shift of the band gap quantified by Varshni's law, the peak wavelength emission reduces to 255 nm at 80 K [196]. Monochromatic CL images acquired at

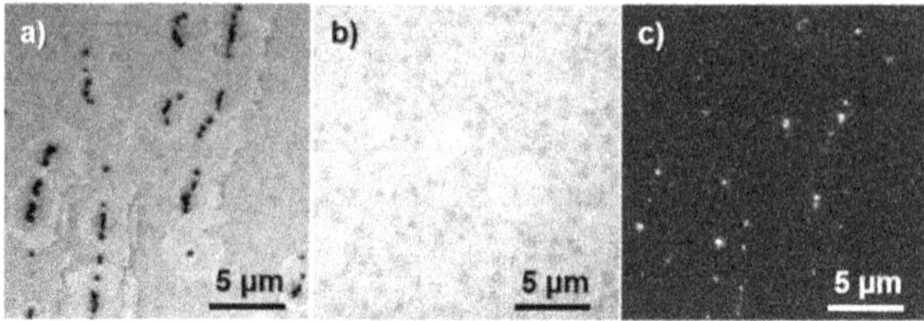

Figure 4.9: Monochromatic CL images (plan-view) acquired at 255 nm of the AlGaN MQW for the **(a)** 0.2° and **(b)** 0.1° offcut sample. **(c)** Monochromatic CL image at 318 nm at the same position of the 0.2° offcut sample as in **(a)**. From reference [134] (first authorship) with permission from Elsevier © 2020.

the peak wavelength of 255 nm are depicted in Fig. 4.9a,b. While for the 0.2° offcut sample dark pits are observed, the luminescence is homogeneous for the 0.1° offcut sample. Some of the dark pits show parasitic luminescence at a longer wavelength of 318 nm, which can be observed by comparing the monochromatic CL images acquired at wavelengths of 255 nm and 318 nm for the 0.2° offcut sample (Fig. 4.9a+c). The pits themselves seem to be aligned in straight lines following the arrangements of the edges of the former step bunches of the 0.2° offcut sample after AlN overgrowth (compare Fig. 4.6c). Obviously, such pits are unwanted and will decrease the overall luminescence performance of the MQW. For the MQW on top of the 0.1° offcut sample the desired homogeneous luminescence is achieved enabled by providing an atomically smooth AlN surface without any step bunches.

For gaining deeper insight into the morphology and formation of the pits and their luminescence at the longer wavelength of 318 nm, cross-sectional STEM investigations were conducted by Anna Mogilatenko and Jonas Weinrich at FBH and the Humboldt University of Berlin. The results are shown in Fig. 4.10. The pits show a V-shape morphology and hence are referred to as V-pits in the following. The formation of the V-pits takes place already deep inside of the AlGaN. In Fig. 4.10a a slightly higher image intensity is observed in the center of the V-pits indicating a potential Ga-rich filling in the center of the V-pits. To further investigate this phenomena of the increased Ga-content inside the V-pits, energy dispersive X-ray (EDX) spectra were acquired on positions inside the V-pits and on the surrounding material. The results are plotted in Fig. 4.10c while the related spatial positions can be found in Fig. 4.10b. The position in the center of the V-pit clearly shows a higher Ga-L and lower Al-K signal in comparison to the position on the surrounding material. Hence, the EDX data is consistent with the hypothesis of Ga-rich filling in the center of the V-pits. Furthermore, the Ga-rich filling

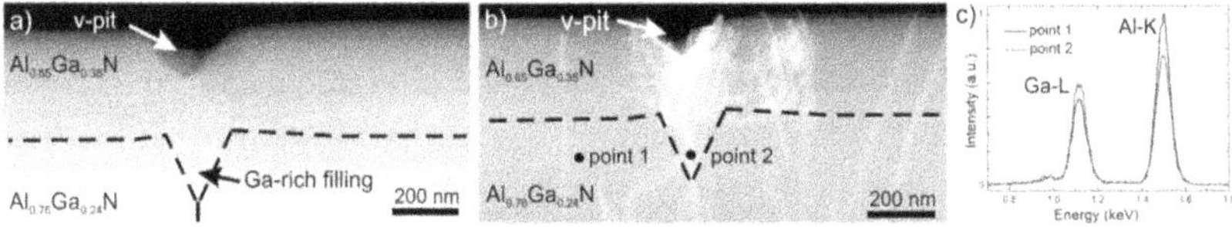

Figure 4.10: **(a)** Cross-sectional STEM image of the AlGaN layer stack for growing the MQW on top of the 0.2° off-cut sample with one of the V-pits. **(b)** ADF STEM image of the identical sample position showing that the V-pit formation occurs above a dislocation line. The dislocation contrast shows that a dislocation lies directly below the V-pit. **(c)** EDX spectra carried out at sample positions inside a V-pit and on the surrounding material as denoted in **(b)**. An increased Ga-content inside the V-pit is verified. This Ga-rich (Al,Ga)N filling most likely results in the dislocation network close to the V-pit seen in **(b)**, since it distinctively alters the strain state. From reference [134] (first authorship) with permission from Elsevier © 2020.

also explains the observed luminescence at 318 nm (Fig. 4.9c) which would correspond to $Al_{0.18}Ga_{0.82}N$. The cause of the Ga-rich filling lies in preferential Ga-incorporation on the semipolar facets of the additional free surfaces inside the V-pits. Preferential Ga-incorporation takes place due to the higher diffusion length of the Ga-adatoms compared to the Al-adatoms during AlGaN growth [192, 197, 198]. This effect can be further strengthened due to reduced compressive strain close to the sidewall facets so that even more Ga instead of Al is incorporated into the crystal lattice inside the V-pits. The underlying cause of the formation of the V-pits can be deduced by taking Fig. 4.10b under consideration. Right below the origin of the V-pit, a TD can be seen. TDs are the preferential points leading to the formation of the pits. However, the density of the V-pits is much smaller compared to the overall TDD. Consequently, not all TD lead to the formation of a related V-pit. TDs with exceptionally large Burgers vectors are most likely the type of TDs leading to an accompanied V-pit. Similar observations have been made in InGaN alloys [199–201]. If the Burgers vector of a TD becomes too large, the formation of free surfaces becomes favourable in terms of energy minimisation of the crystal lattice. As could be seen by DSE (Fig. 4.8c), the TDs tend to line up along step bunches. This increases the probability of dislocation reaction not only leading to annihilation but also dislocation clustering depending on the Burgers vectors of the reacting TDs. Such dislocation clusters are potential candidates for building up large Burgers vector and hence, for the formation of the observed V-pits and would also explain the lining up along the underlying step bunches of the AlN. Despite the presence of free surfaces inside the V-pits, which can represent a certain strain relaxation mechanism, the growth of Ga-rich material inside them leads to further strain relaxation by formation of a broad net of dislocations around the pits (Fig. 4.10b). These defect-rich regions are spatially localized around the pits, but appear directly within the active region. Obviously, such pits will limit the final device performance of an LED by limiting mainly the IQE but also being possible electrical short-cuts leading to increased leakage current [202]. In contrast to that, for the 0.1° offcut sample, the atomically smooth surface leads to homogeneous AlGaN growth and thus, homogeneous luminescence of the AlGaN MQW which is the desired outcome. So besides the demonstration of the improvement by reducing the offcut to 0.1°, it was proven that an atomically smooth surface can be of crucial importance for the realisation of high performance UV LEDs. This might not be valid in all cases, since there are special growth methods for which a rough surface can be beneficial [203].

In conclusion, a suitable AlN template on top of the nanopillar-patterned sapphire could be achieved by reducing the offcut from 0.2° to 0.1° ending up with an atomically smooth surface and decent crystal quality after AlN overgrowth. However, the overgrowth of the nanopillars stays challenging and contains major drawbacks: The temperature of 1380 °C necessary for coalescence is the maximum temperature of the utilised growth reactor and in general relatively high. In the developed growth method it has to be applied for 3 h, which will result in fast attrition of all reactor parts making the method not very cost effective. On the other hand, many of the motivations for growing on top of NPSS instead of using a micropattern introduced at the beginning of this chapter could not be achieved:

- The coalescence thickness could not be reduced and is in the same range as for AlN overgrowth of sapphire stripe patterns in the micrometre range. This results in severe wafer bow which will be detrimental for the processing of final LED devices.

- NPSS was expected to further reduce the TDD by efficient bending of the dislocations towards free surfaces at growth start due to the small pattern dimensions. Although the overgrown nanopillars show a decent crystal quality indicated by a TDD in the range of $1 \times 10^9\,\mathrm{cm}^{-2}$, again it lies in the same range as what can be achieved for overgrown patterns in the micrometre range.

To understand why the TDD stays in the same range as for overgrowing stripe patterns, a cross sectional STEM image is exemplarily shown for the overgrown 0.2° offcut sample in Fig. 4.11. The measurement was again conducted by Anna Mogilatenko and Jonas Weinrich at FBH and the Humboldt University of Berlin. In Fig. 4.11a the expected dislocation bending towards the free surfaces of the not yet coalesced AlN clearly takes

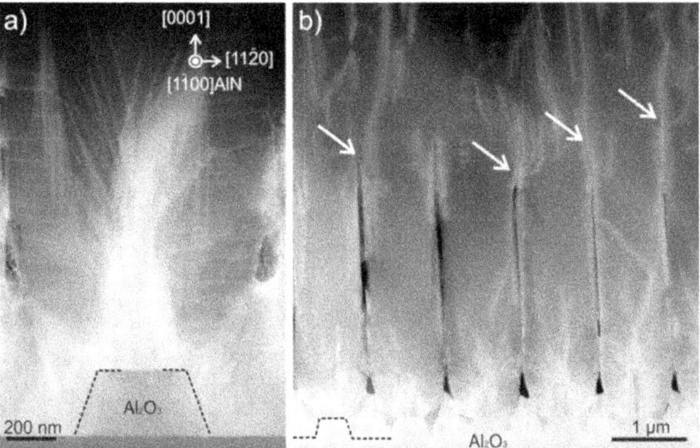

Figure 4.11: Cross-sectional ADF STEM images of the overgrown 0.2° offcut sample. **(a)** At growth start dislocation bending occurs so that the dislocations end at the free surfaces of the not yet coalesced AlN. **(b)** The overall image of the whole layer shows that many dislocations are generated at coalescence (marked by the white arrows). They tend to form networks which spread laterally during subsequent growth leading to a strong increase of the final TDD.

place. This results in almost no left-over dislocations in the independently growing AlN grains on each nanopillar. However, the situation dramatically changes as soon as coalescence starts to develop (Fig. 4.11b): A lot of new TDs are generated at coalescence points and they even form dislocation networks during subsequent growth so that they tend to spread laterally across the whole layer. These networks of TDs most likely form, because the AlN grains on each nanopillar are slightly twisted and/or tilted to each other due to their independent nucleation. As soon as they coalesce, low angular grain boundaries form with accompanied networks of TDs to compensate the tilt and/or twist. During further growth of the smoothing step, the new TDs spread across the layer so that the underlying pattern is not observed in the final defect distribution made visible by ECCI and DSE (Fig. 4.8). Furthermore, the cross-sectional TEM image in Fig. 4.11b again shows the inhibited coalescence: Although the different grains initially show an effective lateral growth so that they almost coalesce already after several hundreds of nanometres, full coalescence only happens after growth of several micrometres.

All these outcomes lead to the conclusion that the nanopillar pattern might not be the perfect candidate for AlN overgrowth and subsequent usage as UV LED template. Thus, in the next section of this chapter, the inverse pattern represented by nanoholes will be investigated in terms of AlN overgrowth and further usage as UV LED template. The initial motivation for choosing nanopillars instead of nanoholes was strain management

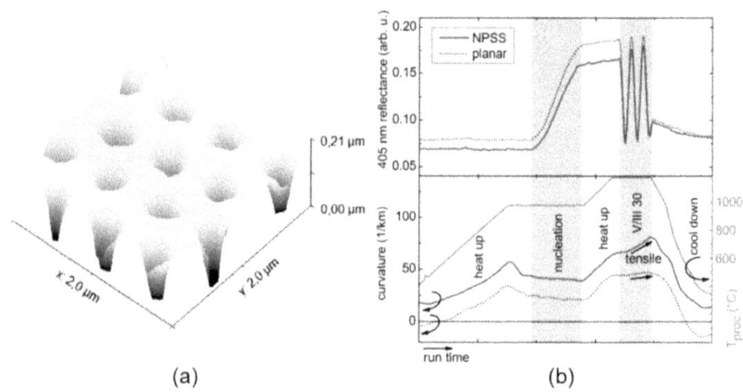

Figure 4.12: **(a)** 3-dimensional AFM topogram of the nanohole NPSS with 0.2° offcut fabricated by nano-imprint lithography and subsequent dry etching. The hexagonally aligned nanoholes have 600 nm pitch, 220 nm hole diameter and 200 nm hole depth. **(b)** In-situ measurement including 405 nm reflectance, process temperature and curvature against runtime for growth of a 300 nm AlN starting layer on top of the sapphire nanoholes as well as on planar sapphire as reference. For growth on the sapphire nanoholes tensile strain develops shortly after nucleation indicated by the increasing curvature. From reference [135] (second authorship) with permission of John Wiley and Sons © 2020.

during overgrowth. Due to the continuous growth surface in case of the nanoholes, tensile strain is expected to be incorporated directly after growth start in contrast to growth on non-continuous patterns like the nanopillars or also the 1-dimensional stripes. This issue will be the major challenge to tackle on the way to a fully coalesced and crack-free AlN layer on top of sapphire nanoholes.

4.3 Growth on sapphire nanoholes

For the following investigations of overgrowing sapphire nanoholes with AlN by MOVPE, NPSS of different fabrication methods and varying pattern dimensions have been used. The first batch NPSS with a sapphire offcut of 0.2° towards the m direction was fabricated together with a project partner in the Advanced UV for Life Consortium using nano-imprint lithography and subsequent dry etching. A 3-dimensional AFM topogram of the nanohole pattern is depicted in Fig. 4.12a. The pattern consists of hexagonally aligned holes with a pitch of 600 nm, hole diameter of 220 nm and hole depth of 200 nm. The main problem in overgrowing such a pattern lies in the development of tensile strain during growth. Due to the continuous surface at growth start (in contrast to growth on the nanopillars) tensile growth is expected similarly to growth on planar

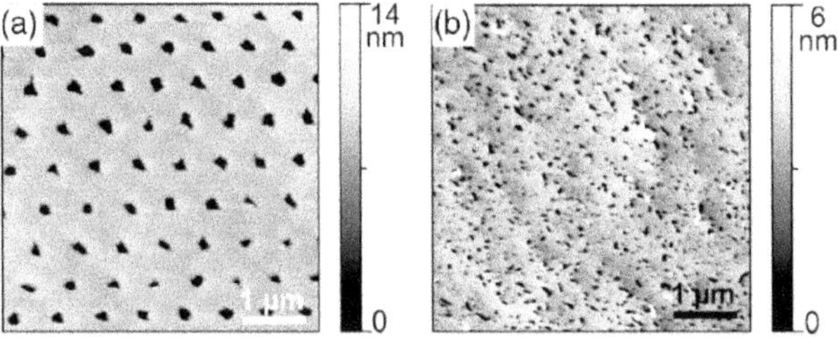

Figure 4.13: AFM images of the 300 nm AlN starting layer grown on top of the **(a)** sapphire nanoholes and **(b)** on planar sapphire. The sapphire nanoholes are not yet coalesced, but the AlN in-between already forms a continuous layer. In contrast to that, for the planar sapphire small pits are present so that the layer is not yet continuous. From reference [135] (second authorship) with permission of John Wiley and Sons © 2020.

sapphire.

To investigate the development of tensile strain during growth, as a starting point a thin AlN layer of 300 nm thickness was grown on the nanohole-patterned sapphire and on planar sapphire as reference simultaneously in the same growth run. After standard thermal cleaning and standard nucleation, the growth temperature was set to 1180 °C and the V/III ratio to 30 for growth of 250 nm AlN. The in-situ measurement of the growth including 405 nm reflectance, process temperature and curvature against runtime is displayed in Fig. 4.12b for growth on the sapphire nanoholes and on planar sapphire. The 405 nm reflectance shows the typical Fabry-Pérot oscillations indicating smooth growth for both samples. The average reflectance on the planar sapphire is higher compared to the AlN on top of the nanoholes, since the not yet coalesced holes lead to scattering of the 405 nm light. An increase in the average reflectance can be observed for the nanohole sample during growth of the 250 nm AlN due to lateral overgrowth of the nanoholes. The curvatures of both samples after nucleation behave very different: While the curvature stays almost constant during growth of the 250 nm AlN on the planar sapphire, it starts to linearly increase shortly after nucleation for the nanohole sample indicating the build-up of tensile strain. This observation is counter-intuitive, since the effective surface area of the nanohole-patterned sapphire is reduced in comparison to the planar sapphire and hence, should lead to reduced tensile strain during growth. However, the opposite is the case.

Considering AFM images of the surface of the samples after growth of the 300 nm AlN (Fig. 4.13), this effect can be understood better. The AlN on the planar sapphire has many small pits. The AlN on the sapphire nanoholes only shows the regularly aligned nanoholes, which are not yet coalesced, while no small pits similar to the AlN on the planar sapphire are observed. As mentioned in the previous chapter 3.2, the polarity of the nucleation islands is a mixture of Al- and N-polarity. The ratio of this mixture strongly depends on the oxygen concentration of the surrounding reactor atmosphere during nucleation [81]. The higher the oxygen concentration the more AlN nucleation islands show Al-polarity. The Al-polar nucleation islands are favoured by the subsequent growth conditions leading to overgrowth of the N-polar domains until the full layer of pure Al-polarity forms. If more N-polar domains are present, coalescence of the Al-polar nucleation islands is retarded. The small pits observed for the AlN on planar sapphire are most likely due to the not yet coalesced Al-polar nucleation islands. Since no closed layer has formed yet also the curvature stays constant, because no tensile strain builds up. For the nanohole sample the AlN between the nanoholes already forms a closed layer leading to tensile growth and thus, the increase in the curvature, although the nanoholes are not yet coalesced. The question remains why the AlN between the nanoholes already forms a closed layer in contrast to the planar sapphire for which the Al-polar nucleation islands are not yet coalesced. An explanation for that could be a higher oxygen concentration during nucleation in the surrounding reactor atmosphere for growth on the sapphire nanoholes compared to growth on the planar sapphire [81]. This would lead to a higher amount of Al-polar nucleation islands. A potential candidate for the additional oxygen source could be the nanohole pattern itself. The non c-plane side facets of the sapphire are less stable during heating up and thermal cleaning under hydrogen [204]. Hence, small parts of the sapphire are etched by the hydrogen leading to an increased oxygen concentration immediately before nucleation and at the growth start before the sapphire is completely covered by AlN.

In conclusion, this experiment shows that the expected problem of build-up of tensile strain during overgrowth of sapphire nanoholes is even more severe than for growing on planar sapphire. This will lead to macroscopic cracking of the layer already at a thin layer thickness. For the 300 nm AlN starting layer no cracks were observed in the light microscope. However, cracks were observed before coalescence if growth under the same conditions was simply continued. To end up with a coalesced surface without any cracking of the layer, the strain of the layer has to be managed. One method of strain management, which was introduced in chapter 3.2, is the approach to cause roughening of the layer at growth start and subsequent smoothing by certain growth conditions. However, this is much harder to achieve for growth on the sapphire nanoholes than for growth on planar sapphire, since the area between the nanoholes offers only a

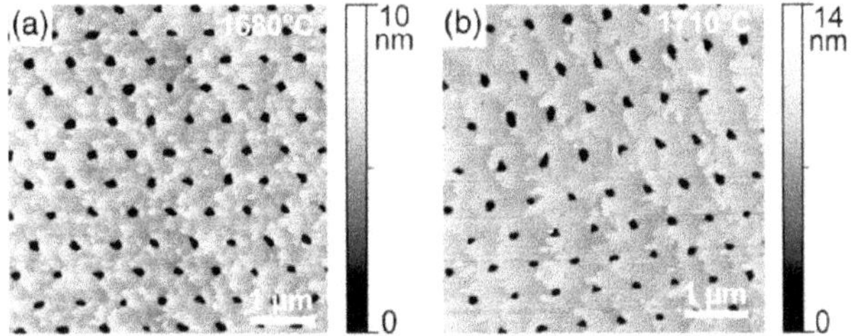

Figure 4.14: AFM images of the 300 nm AlN starting layer grown on top of the sapphire nanoholes after HTA at **(a)** 1680 °C and **(b)** 1710 °C for one hour. No surface damage is observed so that the layers appear suitable for further AlN overgrowth. From reference [135] (second authorship) with permission of John Wiley and Sons © 2020.

very limited amount of sapphire *c*-plane to grow on. That makes it very difficult to find appropriate growth conditions for roughening so that a subsequent smoothing is achieved rather than the layer just staying rough. Another potential method for strain management is the HTA investigated in chapter 3.2.

One of the outcomes of the method of HTA was the change of the strain of an annealed layer to more compressive values translating into a reduction of the in-plane lattice constant. If this could be achieved for AlN on the sapphire nanoholes, the problem of tensile growth should be suppressed similar to the overgrowth of the annealed layer shown in chapter 3.2 in Fig. 3.11. Additionally, HTA reduces the TDD which is of course also beneficial. To meet the challenge of tensile growth on nanohole NPSS, 300 nm AlN starting layers on top of the sapphire nanoholes were annealed in the annealing furnace at FBH. Two annealing temperatures (1680 °C and 1710 °C) were chosen for comparison while the annealing time was set to 1 h. AFM images of the surface after annealing can be seen in Fig. 4.14. For both annealing temperatures, no decomposition of the material or other kinds of surface damage could be observed. Similar to the annealed layers in the previous chapter, small step bunches tend to form between the nanoholes. Overall, the layers look suitable for further AlN overgrowth in terms of surface morphology. The material quality before and after annealing was checked by XRD. The XRC-FWHM of the 0002 and 10-12 reflections and the estimated TDDs for the 300 nm AlN starting layers with annealing at the different temperatures and without annealing are found in Tab. 4.3. The obtained values suggest a distinct reduction of about one order of magnitude of the TDD after annealing similar to the experiments presented in chapter 3.2. The difference between the two layers annealed at the different temperatures of 1680 °C and 1710 °C is negligible in terms of material quality. Unfortunately, it was

Table 4.3: XRC-FWHM of the 0002 and 10-12 reflections measured by XRD and corresponding TDDs estimated by calculation for the 300 nm AlN starting layers and fully overgrown 800 nm AlN layers grown on nanoholes with intermediate annealing at 1680 °C and 1710 °C for one hour and without (w/o) annealing.

Layer description	0002	10-12	TDD
300 nm starting layer (w/o HTA)	200″	820″	$9 \times 10^{9}\,\mathrm{cm}^{-2}$
300 nm starting layer (1680 °C)	100″	410″	$2 \times 10^{9}\,\mathrm{cm}^{-2}$
300 nm starting layer (1710 °C)	90″	410″	$2 \times 10^{9}\,\mathrm{cm}^{-2}$
800 nm coalesced layer (w/o HTA)	300″	1650″	$3.2 \times 10^{10}\,\mathrm{cm}^{-2}$
800 nm coalesced layer (1680 °C)	100″	450″	$2 \times 10^{9}\,\mathrm{cm}^{-2}$
800 nm coalesced layer (1710 °C)	90″	430″	$2 \times 10^{9}\,\mathrm{cm}^{-2}$

not possible to determine the strain of the AlN starting layers by XRD, since the layers consist of only 300 nm AlN thickness so that the signal-to-noise-ratio was too low to obtain reliable results. However, the strain of the layer is visible indirectly in the curvature measurement during subsequent AlN overgrowth.

For subsequent overgrowth the aim is to coalesce the layer and end up with a surface as smooth as possible. For efficient lateral growth, the highest growth temperature of 1380 °C is chosen analogous to overgrowing the sapphire nanopillars in chapter 4.2. However, for the usual V/III ratio of 30 and the present offcut of 0.2°, it is expected that problematically high step bunches will form on the surface similar to growth on nanopillars. Hence, the V/III ratio is set to a relatively high value of 810. For that V/III ratio AlN grown on the sapphire nanopillars fails to coalesce [97]. However, in case of the nanoholes coalescence should be easier to achieve due to the continuous surface at growth start and the smaller coalescence fronts compared with the overgrowth of nanopillars. The in-situ measurement including 405 nm reflectance, process temperature and curvature against runtime is shown in Fig. 4.15 for overgrowing the two starting layers annealed at different temperatures and the starting layer without annealing as reference. All samples were overgrown simultaneously in one single growth run. The 405 nm reflectance indicates smooth growth with the typical Fabry-Pérot oscillations for all samples and their average reflectance tends to increase due to the lateral overgrowth of the nanoholes. However, this increase can only slightly be seen, since the nanoholes are relatively small in comparison to the reflectance wavelength of 405 nm. After growth of 500 nm the growth was stopped, because the constant average reflectance suggests coalescence of the samples. The curvature shows a strong linear increase over time for the sample without intermediate HTA. In contrast to that, for both annealed samples the curvature stays almost constant for the whole growth. This observation is consistent

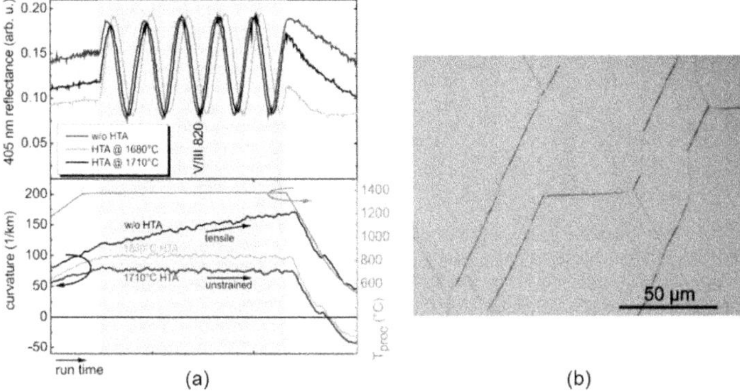

Figure 4.15: (a) In-situ measurement including 405 nm reflectance, process temperature and curvature against runtime for growth of 500 nm AlN on top of the 300 nm starting layers with and without intermediate HTA. The HTA samples grow strain free in contrast to tensile growth for the sample without HTA. **(b)** Light microscope image of the 800 nm layer without intermediate HTA showing macroscopic cracking of the surface. From reference [135] (second authorship) with permission of John Wiley and Sons © 2020.

with the results in chapter 3.2, in particular Fig. 3.11. There, it was observed that the HTA step leads to a reduced in-plane lattice constant and thus, compressive rather than tensile growth. The in-situ curvature in Fig. 3.11 even showed a linear decrease translating to the build-up of compressive strain. The growth was performed at 1180 °C. In the current case, the growth is performed at 1380 °C. The higher growth temperature leads to a desired smaller in-plane lattice constant of the growing AlN so that the growth now becomes strain-free rather than compressive. The curvature values obtained after cool down are in the range of $-40\,\mathrm{km}^{-1}$ for the two layers with intermediate annealing step and $40\,\mathrm{km}^{-1}$ for the layer without intermediate annealing. All these values are relatively small which is a good result considering processing of potential UV LEDs.

The macroscopic surface morphology of all three samples was investigated by light microscopy imaging. While cracks were observed across the whole wafer for the sample without intermediate annealing (Fig. 4.15b), no cracks were found for the samples with the intermediate HTA step. The difference is obviously related to the strain of the layers during overgrowth. Additionally, the microscopic surface morphology of all samples was investigated by AFM. The corresponding images are depicted in Fig. 4.16. All samples show full coalescence of the nanoholes with a step-bunched AlN surface. The step bunches are of similar height for all three samples in the range of 5 nm. Although an atomically smooth surface with bilayer steps is preferential, the small step bunches

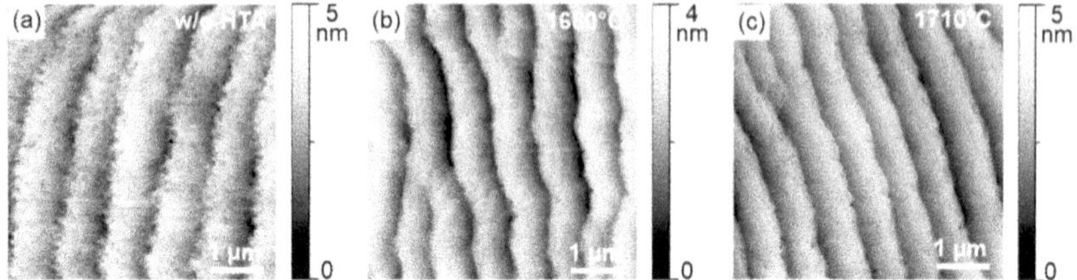

Figure 4.16: AFM images of the overgrown 800 nm AlN layers grown on top of the sapphire nanoholes **(a)** without (w/o) intermediate HTA and with intermediate HTA at **(b)** 1680 °C and **(c)** 1710 °C for one hour. All samples show step bunches on the surface with step height of about 5 nm. From reference [135] (second authorship) with permission of John Wiley and Sons © 2020.

should not be an issue for further epitaxial growth of device layers, since the height of the step bunches is much smaller than the critical value of about 10 nm. In conclusion, by introducing the intermediate HTA step, a fully coalesced AlN layer without any macroscopic cracks was achieved on top of the sapphire nanoholes. Hence, the strain management was successful. To check the final material quality, the XRC-FWHM after overgrowth were measured and the corresponding TDDs were estimated by calculation (Tab. 4.3). The overgrown layer without the intermediate annealing step shows higher XRC-FWHM after overgrowth than before. This can be explained by broadening of the XRC-FWHM due to layer cracking. For both layers with intermediate annealing step, the improved material quality due to annealing could be conserved. They both show the same estimated TDD of $2 \times 10^9\,\mathrm{cm}^{-2}$ for a layer thickness of only 800 nm AlN. Considering the underlying nanopattern which potentially improves the light extraction efficiency of a final LED device, the achieved AlN template is a promising candidate as a base layer for fabrication of high performance UV LEDs. In the following, for overgrowing other nanohole patterns, the temperature for the intermediate HTA step is set to 1680°. No beneficial effect was found for the higher temperature of 1710°, while a higher annealing temperature makes decomposition of material and degradation of the interface more likely [180].

In the following, the developed growth method is applied to nanohole-patterned sapphire with different pattern dimensions. In the first part of this chapter (4.1), it was shown that the pattern dimensions have an influence on the potential improvement of the light extraction efficiency of the final UV LED. Thus, it is desired that the developed growth method also works for nanohole patterns with varying pattern dimensions. Additionally, the influence of the pattern dimensions on the final TDD will be investigated. For that purpose, a second batch of nanohole NPSS with varying

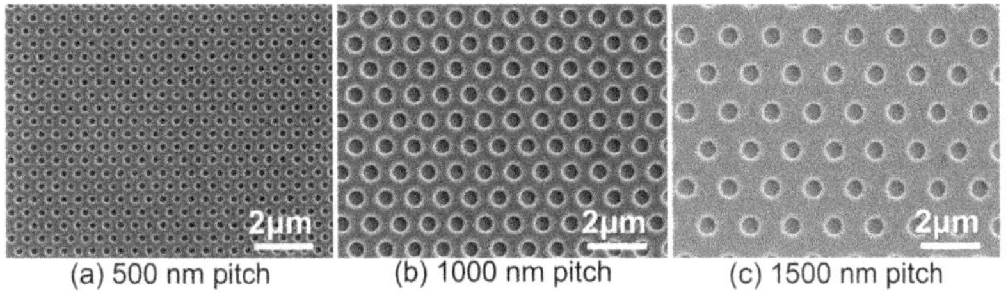

Figure 4.17: Plan view SEM images of nanohole patterns with pitch of **(a)** 500 nm, **(b)** 1000 nm and **(c)** 1500 nm fabricated at the University of Bath by Pierre-Marie Coulon. Hole diameters are 200 nm, 475 nm and 545 nm, respectively, while the hole depth is 500 nm for all samples.

pitch was prepared by Pierre-Marie Coulon at the University of Bath. The fabrication procedure is similar to the fabrication of the sapphire nanopillars applying Displacement Talbot Lithography. With this technique it is easily possible to change the pattern dimensions e.g. pitch and hole diameter [137]. A detailed description of the fabrication process can be found in chapter 2.1. For this batch of wafers the offcut could be selected so that 0.1° to the m direction was chosen to suppress the formation of step bunches. A series of nanohole patterns with three different pitches (500 nm, 1000 nm and 1500 nm) was prepared. SEM images of the patterns are shown in Fig. 4.17. The hole diameter is 200 nm for the 500 nm pitch pattern, 475 nm for the 1000 nm pitch pattern and 545 nm for the 1500 nm pitch pattern. The ratio between hole diameter and pitch was tried to be kept constant at 0.5. However, deviations could not be fully prevented and are due to the explorative fabrication method. The hole depth is 500 nm for all samples.

The samples were overgrown according to the previously developed three-step growth method applying the exact same growth conditions:

1. Growth of a 300 nm AlN starting layer.

2. Intermediate annealing at 1680° for 1 h.

3. Overgrowth until coalescence is reached.

For overgrowth of the annealed starting layers, the point of coalescence is indicated by saturation of the in-situ average 405 nm reflectance. The duration of the overgrowth step is set manually to ensure coalescence of all layers. For better comparison, all samples were overgrown in the same growth run up to the same layer thickness. The in-situ measurement of overgrowing the starting layers after HTA including 405 nm

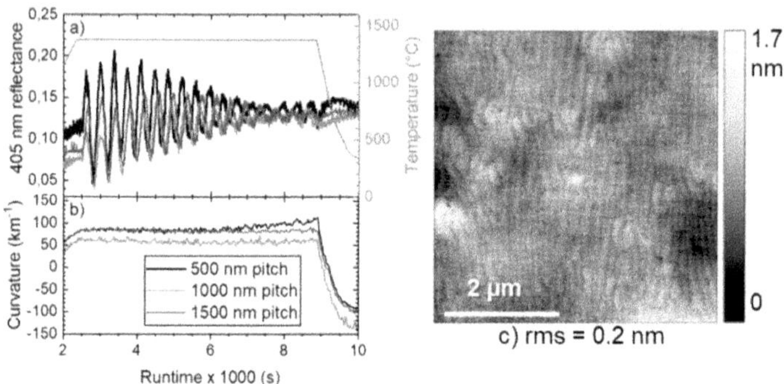

Figure 4.18: (a)+(b) In-situ measurement including 405 nm reflectance, process temperature and curvature against runtime for growth of 1.7 µm AlN on top of the annealed starting layers for underlying nanohole patterns with pitch of 500 nm, 1000 nm and 1500 nm. **(c)** AFM image exemplarily shown for the sample with pattern pitch of 1500 nm. The surface morphology of all pattern pitches show full coalescence and an atomically smooth AlN surface due to the reduced sapphire offcut of 0.1°.

reflectance, process temperature and curvature against runtime is shown in Fig. 4.18. The 405 nm reflectance shows the typical Fabry-Pérot oscillations for all samples. The average reflectance is different for each nanohole pattern and corresponds to their specific nanohole diameter. For all samples the average reflectance increases, since lateral growth occurs at the high growth temperature of 1380°. Coalescence is reached at different points in time during the growth run: For the 500 nm pattern, coalescence is reached after approximately 800 nm total AlN layer thickness, while for the 1000 nm and 1500 nm patterns coalescence is reached after approximately 1600 nm total AlN layer thickness. Due to the very similar hole diameter of the two samples with larger pitch, coalescence occurs at similar points in time. The growth is stopped after 1.7 µm additional AlN growth leading to a total layer thickness of 2 µm including the AlN starting layer. In comparison to growth on the sapphire nanopillars, coalescence of the holes seems to happen much faster even for larger pattern pitches. This meets the expectations due to the continuous surface at growth start and the smaller coalescence fronts. For coalescence of the nanopillars, the independent nucleation on each pillar generates more twist and/or tilt which eventually inhibits coalescence.

The curvature of the samples stays constant for growth on the 1000 nm and 1500 nm pattern during the whole growth. For the sample with the 500 nm pattern the curvature at first stays constant, but then starts to slightly increase after coalescence. Thus, the

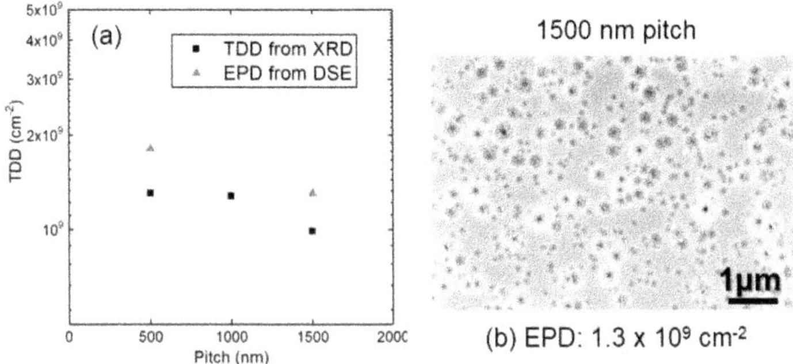

Figure 4.19: (a) TDD estimated from XRC-FWHM and EPD after DSE for the fully overgrown nanohole patterns with different pitches. For the 1000 nm no EPD could be obtained after DSE due to overetching. **(b)** SEM image of the fully overgrown 1500 nm pitch sample after DSE to exemplarily show the determination of the EPD.

intermediate HTA step does not enable endless strain-free or compressive growth, but the AlN growth will eventually switch to a tensile mode again. This happens first for the sample with the 500 nm pattern since coalescence is reached first. No cracks could be observed after overgrowth for all of the samples. This was checked by light microscopy. To investigate the microscopic surface morphology AFM measurements were carried out. The microscopic surface looks identical for all layers and is exemplarily shown for the sample with pattern pitch of 1500 nm in Fig. 4.18b. Full coalescence is achieved and the surface is atomically smooth so that the typical AlN bilayer steps are observed. The reduced offcut of 0.1° successfully shifted the growth mode to step-flow growth while still attaining coalescence. For determining the material quality by estimating the TDD, measurements of the XRC-FWHM of the 0002 and 10-12 reflections were performed. Additionally, the EPD was counted after performing DSE. The obtained values of both methods are plotted against the pattern pitches in Fig. 4.19a. An SEM image of the etched surface for counting the EPD is exemplarily shown for the sample with pattern pitch of 1500 nm in Fig. 4.19b. Unfortunately, the surface of the sample with pattern pitch of 1000 nm was overetched so that no EPD could be determined. The first thing to note is that in case of the nanoholes the EPDs are very similar to the TDDs estimated from the XRC-FWHM. Hence, the wingtilt assumed for growth on nanopillars does not significantly broaden the XRC-FWHM in case of growth on nanoholes. This is most likely due to the reduced area of the wing regions for overgrowing nanoholes instead of nanopillars. Secondly, the lowest TDD seems to be reached for the biggest pattern pitch of 1500 nm with an EPD of $1.3 \times 10^9\,\mathrm{cm}^{-2}$ and TDD estimated by XRD of $1 \times 10^9\,\mathrm{cm}^{-2}$. An explanation for that would be that for the larger holes the total amount of coalescence fronts becomes smaller in comparison to the smaller holes assuming the aspect ratio is kept constant [100]. Since coalescence fronts lead to generation of new dislocations as was already seen in chapter 4.2 in case of the nanopillars, the larger nanoholes lead to a reduction of the TDD. However, the effect is not very pronounced: The sample with pattern pitch of 1000 nm showed a similar hole diameter compared to the sample with pattern pitch of 1500 nm, but the obtained values for EPD and TDD are more similar to the sample with pattern pitch of 500 nm. Nonetheless, the absolute values obtained for all samples show decent material quality with a TDD in the range of $1 - 2 \cdot 10^9\,\mathrm{cm}^{-2}$. Overall, the applied three-step growth method behaves stable under variations of the pattern dimensions.

In the last part of this chapter, AlN growth on nanohole-patterned sapphire substrates bought from a commercial supplier is investigated. The availability of a reliable commercial supplier is a crucial benchmark towards the aim of realising a stable technological process which in principal can be scaled up. The commercial supplier offers nanohole NPSS at a prize of only 3-fold higher than planar sapphire. The fabrication procedure of the nanohole NPSS is unknown, but most likely applies nano-imprint lithography, since this technique is well established and cost-effective for large scale production. The offcut is fixed to 0.2° towards the *m* direction. A plan view SEM image of the pattern is shown in Fig. 4.20a. It can be seen that big parts of the sapphire between the holes are not of pure *c*-plane indicated by the different contrast in the SEM image. The sample is overgrown following the same three-step growth method with overgrowth up to a final layer thickness of 2.2 µm. An AFM image of the surface after complete overgrowth is shown in Fig. 4.20b. Despite the small amount of remaining *c*-facet observed in the SEM image of the patterned sample before overgrowth, full coalescence is reached. This again proves the stability of the developed growth process with regard to different nanohole patterns. Similar to the first batch of wafers investigated in the beginning of this chapter the surface shows step bunches of around 5 nm height due to the offcut of 0.2°. The TDD estimated from the acquired XRC-FWHM of the 0002 and 10-12 reflection is $1.3 \times 10^9\,\mathrm{cm}^{-2}$. In conclusion, all the different batches of nanohole NPSS behave very similar when applying the developed three-step growth method. In terms of reproducibility of the growth method this is a promising result.

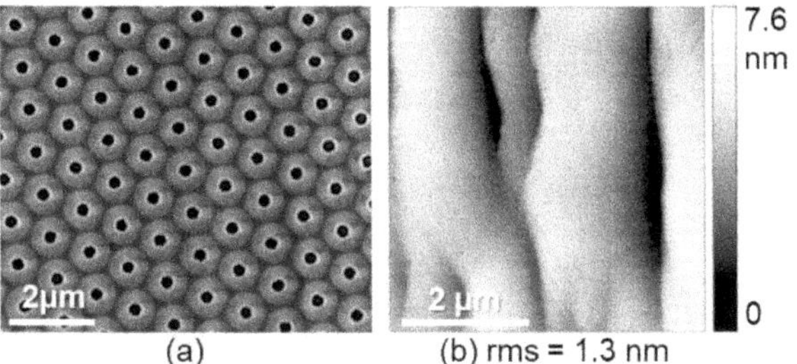

Figure 4.20: **(a)** SEM image in plan view of the nanohole-patterned NPSS bought from the commercial supplier. The pattern pitch is 1000 nm, hole diameter 370 nm and hole depth 520 nm. **(b)** AFM image of the sample showed in **(a)** after applying the three-step growth method described earlier in this section. Full coalescence is achieved while the surface shows step bunches of about 5 nm height due to the sapphire offcut of 0.2°.

However, the formation of step bunches and also the very high growth temperature of 1380 °C is still unwanted. To address both of these problems, further growth experiments were conducted with the substrates from the commercial supplier. The idea was to reduce the growth temperature during overgrowth of the annealed starting layers. Furthermore, by decreasing the temperature the detrimental effect of the high growth temperature to the reactor parts can be reduced. The high growth temperature was taken from the experience of overgrowing nanopillars in chapter 4.2. However, in the present chapter it was now shown that coalescence of nanoholes appear to happen much easier compared to the case of nanopillars. For the following growth experiment the well established growth conditions of 1180 °C and V/III ratio of 30 for inducing step-flow growth with a sapphire offcut of 0.2° are chosen. The in-situ measurement of the overgrowth of annealed 300 nm AlN starting layers including 405 nm reflectance, process temperature and curvature is depicted in Fig. 4.21. After standard thermal cleaning, the growth temperature is set to 1180 °C and V/III ratio of 30. The reflectance shows the typical Fabry-Pérot oscillations with average reflectance increasing with runtime until the average reflectance saturates at a value of about 0.14 indicating coalescence of the layer. The lateral growth does not seem to be suppressed by reducing the growth temperature from 1380 °C down to 1180 °C. Coalescence happens at a similar point in time compared with overgrowth at the higher temperature of 1380 °C. This again shows the advantage of the nanohole pattern with the continuous growth surface at growth start. In this case, it becomes easier to maintain the AlN c-facet and achieve subsequent

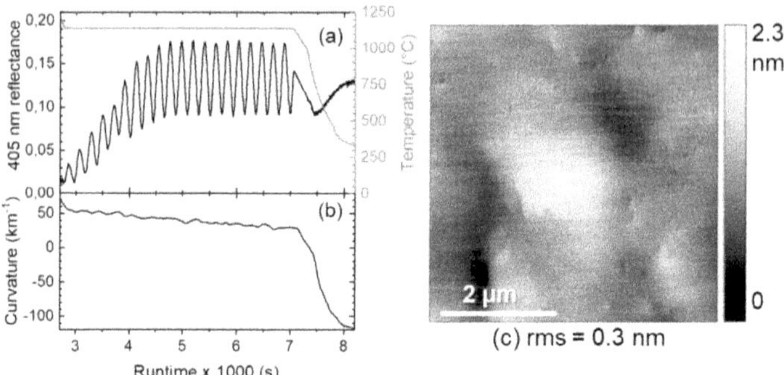

Figure 4.21: (a)+(b) In-situ measurement including 405 nm reflectance process, temperature and curvature against runtime for growth of 1.9 µm AlN on top of the nanohole NPSS presented in Fig. 4.20a. Although the growth temperature was significantly reduced to induce step-flow growth, the layer coalesces. **(c)** AFM image of the sample overgrown in (a) proving that full coalescence and an atomically smooth surface was achieved by reducing the overgrowth temperature down to 1180 °C.

coalescence in comparison to growth on nanopillars for which the high temperature is necessary [97].

The microscopic surface was checked by AFM. The corresponding image is shown in Fig. 4.21b. Full coalescence is proven and the morphology shows an atomically smooth surface with bilayer steps as expected from the applied growth conditions. The material quality was investigated by measuring the XRC-FWHM of the 0002 and 10-12 reflections. The 0002 showed a FWHM of 300″ and the 10-12 380″, which translates into an estimated TDD of $1.3 \times 10^9\,\mathrm{cm}^{-2}$. These are equal results compared with the samples overgrown at the high temperature. In conclusion, the established three-step growth method could be adapted by reducing the temperature of the final overgrowth step to end up with an atomically smooth surface and prevent the reactor parts from heat damage. The obtained AlN templates are promising for usage as UV LED base layers. This will be investigated in the final part of this work.

Summary In this chapter AlN growth on NPSS and the potential benefit of using such substrates was investigated. Wave optical simulations showed that an AlN template with implemented nanopattern at the AlN/sapphire interface has the potential of increasing the light extraction efficiency for UV LEDs fabricated on top. However, this only holds, if the light travels through the layer system multiple times which can be realised for instance by a transparent p-side of the UV LED heterostructure combined with a reflective contact material. For a nanohole pattern with pitch of 1000 nm, hole diameter of 500 nm, hole depth of 250 nm and an assumed effective reflectivity of the p-side of 0.8, the simulation showed an increase in the transmittance through the AlN/sapphire interface by 77 %. The effect is due to a scattering induced softening of the internal total reflection. This results in a non-zero probability for transmittance through the AlN/sapphire interface above the critical angle for total reflection. Differently shaped nanopatterns (nanoholes, nanopillars or nanofrustums) behave similar while a bigger pattern pitch leads to an enhanced softening of the total internal reflection. Growth experiments were performed on both nanopillars and nanoholes to realise suitable AlN templates on NPSS. For growth on nanopillars, incorporation of tensile strain followed by layer cracking is not an issue due to the non-continuous surface at growth start. On the other hand, the large coalescence fronts lead to inhibited coalescence which is achieved only after approx. 4 µm AlN growth at very high temperature of 1380 °C. This favours the formation of high step bunches for the standard sapphire offcut of 0.2° which can not be smoothed. By reducing the offcut to 0.1°, an atomically smooth surface could be achieved after additional approx. 2.5 µm AlN growth in step-flow growth mode after coalescence ending with a relatively thick layer of almost 7 µm. Homogeneous luminescence of subsequently grown MQW was demonstrated on nanopillar NPSS with 0.1° offcut instead of pitting observed on the 0.2° nanopillar NPSS due to the step bunches on the AlN surface. For overgrowing sapphire nanoholes, strain management was necessary to avoid layer cracking before coalescence, since the continuous growth surface leads to a constant build-up of tensile strain. However, by applying HTA to a 300 nm AlN starting layer on top of the nanohole NPSS, compressive strain could be triggered which led to suppression of tensile strain during overgrowth. In that way, HTA was used as a tool to reach for coalescence while avoiding layer cracking. The growth method using the intermediate HTA step was successfully demonstrated on nanohole NPSS of different pattern dimensions with the conclusion that it can be considered as a stable and reproducible process. The growth temperature for the final overgrowth step was reduced to 1180 °C instead of the 1380 °C for growth on the nanopillars. This helps to reduce the wear of the reactor parts. Also, compared to the nanopillars, the coalescence thickness was decreased from 4 µm down to approx. 2 µm for equal pattern pitches of 1000 nm by use of nanoholes. The best TDD achieved was $1 \times 10^9\,\mathrm{cm}^{-2}$ for both the nanoholes and the nanopillars. Although there is efficient dislocation bending towards free surfaces at growth start, many new dislocations are generated at coalescence. The full thickness of the AlN on top of the NPSS was approx. 7 µm in case of the nanopillars instead of only 2 µm for the nanoholes.

In conclusion, it was shown that the implementation of a nanopatterned AlN/interface has a great potential for improving the light extraction efficiency of UVC LEDs grown on top. Furthermore, AlN overgrowth of nanopillar and nanohole NPSS was demonstrated. On both patterns, AlN of atomically smooth surface and decent material quality with TDDs of approx. $1 \times 10^9\,\mathrm{cm}^{-2}$ were achieved. In the next chapter, the performance of UVC LEDs on the developed AlN templates will be evaluated.

CHAPTER 5

UVC LED performance on the developed AlN templates

As a final step of this work, the aim is to evaluate the suitability and quality of the AlN templates developed in chapters 3 and 4 for being used as UV LED base layers. For that purpose, AlGaN based LED heterostructures were grown on top of the AlN templates at the Technical University of Berlin by Norman Susilo. The LED heterostructures were grown in a close coupled showerhead reactor with a targeted emission wavelength of 265 nm. This wavelength is important for the most popular UV LED application of water purification [13]. Furthermore, for an emission wavelength of 265 nm, the Al-content of the AlGaN heterostructure will be relatively high enabling pseudomorphic growth up to the active region. If full pseudomorphic growth is achieved, the TDD of the AlN template will be conserved. In that way, the relation between the luminescence performance of the AlGaN MQW and the quality of the underlying AlN template is more straightforward, since no relaxation processes interfere. The main goal is to demonstrate luminescence at 265 nm of the LED heterostructures grown on top of the AlN templates to prove their general suitability to work as UV LED base layers. Additionally, the gain of the emission power will be evaluated quantitatively by comparing the $L-I$ curves of different AlN templates with the same LED heterostructure on top. For a quantitative comparison of the $L-I$ curves the AlN templates have to be overgrown simultaneously in the same growth run, since run-to-run reproducibility especially over long periods of time can not be guaranteed. There are numerous sources for instabilities over time, e.g. temperature drift, change of the impurity concentration in the reactor atmosphere, lifetime of the reactor parts etc. Hence, an overall comparison of the different AlN templates developed over time and overgrown by LED heterostructures in different growth runs is not intended. Instead, certain comparisons in single growth runs are designed to present the improvement gained by AlN templates developed in this work.

HTA of MOVPE grown AlN on planar sapphire as presented in chapter 3 is the first template technology which will be investigated. A 900 nm AlN layer was prepared and annealed at 1710 °C for 5 h, since these annealing parameters lead to the lowest TDD. For the layer after annealing XRC-FWHM of 65″ for the 0002 reflection and 262″ for the 10-12 reflection were determined, which translates to a TDD of $8 \times 10^8\,\mathrm{cm}^{-2}$ estimated by calculation. For comparison, 1.5 µm AlN on planar sapphire without any post annealing was prepared following the growth process shown in chapter 3.2 in Fig. 3.4. The TDD of the 1.5 µm AlN template was estimated to be $3.9 \times 10^9\,\mathrm{cm}^{-2}$ by calculation from the XRC-FWHM of 64″ for the 0002 reflection and 575″ for the 10-12 reflection. Both AlN templates, the HTA planar and planar, were simultaneously overgrown in one single growth run by an LED heterostructure with targeted emission wavelength at 265 nm as mentioned above. Details of the growth are reported in reference [133]. For on-wafer electroluminescence (EL) measurements conducted by Norman Susilo at the Technical University of Berlin, indium dots were used as contact material. Consequently, the LED results presented in this work are not of fully processed UV LED chips. However, this so called on-wafer In-dot quick test usually works as a good estimation for the potential emission power of the final UV LED chips. The output power of the emitted light in dependence of the applied current ($L - I$ curve) and the emission spectrum at a fixed current of 20 mA was measured for both samples. The results are plotted in Fig 5.1. Regarding the $L - I$ curves in Fig. 5.1a it can be seen that for the HTA planar template the emission power is higher than for the planar template. At 20 mA an emission power of 0.58 mW is reached for the HTA planar template in contrast to only 0.13 mW for the planar template. This more than 4-fold increase is attributed to the reduced TDD and hence, an increased internal quantum efficiency [65]. The emission spectra (Fig. 5.1b) show that main emission occurs at a wavelength of approx. 265 nm for both samples. For the HTA planar template, another emission peak is observed at a wavelength of 334 nm. The emission power of the second peak lies two orders of magnitude below the main emission peak. The second peak might be due to local relaxation effects possibly resulting in formation of hillocks with higher Ga-incorporation at the edges [205]. The annealed AlN layers show a reduced in-plane lattice constant compared to non-annealed AlN for which the AlGaN growth

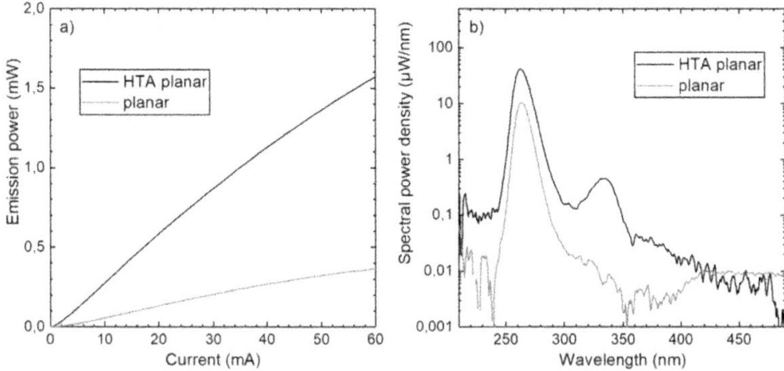

Figure 5.1: **(a)** $L-I$ curves and **(b)** emission spectra at a current of 20 mA for UVC LED heterostructures grown on top of the HTA planar and planar template. The HTA planar template of 900 nm thickness was annealed at 1710 °C for 5 h while the planar template 1.5 µm thickness was not annealed.

conditions were originally optimised.

AlN grown on the nanopillar NPSS with different offcut as described in chapter 4.2 will be investigated in the following template comparison. TDDs of $9 \times 10^8\,\mathrm{cm}^{-2}$ for the 0.2° offcut sample and $1 \times 10^9\,\mathrm{cm}^{-2}$ for the 0.1° offcut sample were determined by defect selective etching. Luminescence of subsequently grown AlGaN MQW on top of these templates was already checked by CL measurements in chapter 4.2. However, there was no quantification of the difference in luminescence. For that purpose, LED heterostructures with a targeted wavelength of 265 nm were grown on top of both templates according to reference [134]. The EL measurements including $L-I$ curves and emission spectra at 20 mA are depicted in Fig. 5.2. The emission power on the NPSS with 0.1° offcut is higher than on the NPSS with 0.2° offcut. At 20 mA an emission power of 0.52 mW is achieved for the 0.1° offcut sample compared to 0.1 mW for the 0.2° offcut sample. This results in a more than 5-fold increase. The less efficient luminescence of the 0.2° offcut sample is ascribed to the pit formation due to the step bunching of the AlN surface (compare Fig. 4.9). No pits were observed for the 0.1° offcut sample. The emission spectra show the main emission peak at a wavelength of approx. 265 nm for both samples. However, for the 0.2° offcut sample an additional peak at 326 nm with a spectral power density 50 times lower than the main emission peak is observed. The wavelength of the additional peak fits well to the luminescence of the Ga-rich regions inside the pits determined by CL (see Fig. 4.9c), which emit

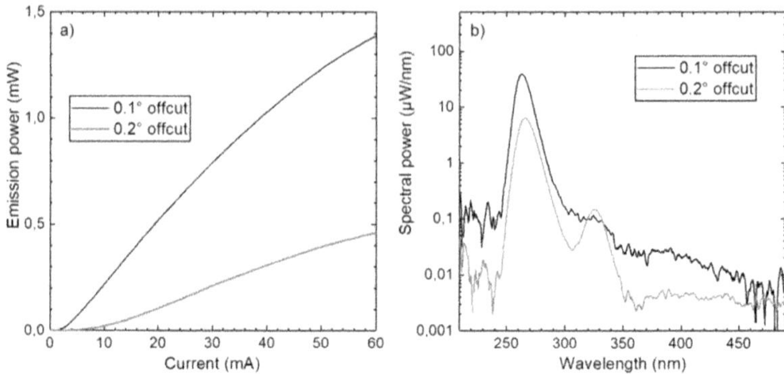

Figure 5.2: (a) $L-I$ curves and **(b)** emission spectra at a current of 20 mA for UVC LED heterostructures grown on top of the AlN templates on the nanopillar NPSS with 0.1° and 0.2° offcut and a pattern pitch of 1 μm.

at 318 nm at a temperature of 80 K. At room temperature this should result in a wavelength of approx. 326 nm.

For evaluating the AlN templates on top of the nanohole NPSS presented in chapter 4.3 an experiment was designed to assess the difference in emission power which is due to the implemented nanopatterned AlN/sapphire interface. In chapter 4.1, wave optical simulations predicted a gain in light extraction efficiency for UV LED heterostructures with an implemented nanopatterned AlN/sapphire interface in case of a non-zero effective reflectivity of the p-side. For a fully absorbing p-side no gain is expected. The challenge to experimentally prove this hypothesis is that the emission power at a fixed current is always dependent on the product of the internal quantum efficiency and the light extraction efficiency. Thus, to assess the light extraction efficiency experimentally the LED heterostructures need to have similar IQEs. The IQE is the product of the injection efficiency and the radiative efficiency. The radiative efficiency is driven by the TDD in the active region of the LED heterostructure, since TDs act as non-radiative recombination centers. Hence, AlN templates with similar TDDs have to be provided to obtain LED heterostructures with similar radiative efficiencies under the assumption that the TDDs of the AlN templates are conserved up to the active region. If further the assumption is made that the LED heterostructures simultaneously grown in a single growth run on AlN templates with similar TDDs will exhibit similar injection efficiencies, the IQE will be similar. In that way, differences in the emission power can be attributed to different LEEs due to for instance different AlN/sapphire interfaces. For the AlN template on top of the nanohole NPSS an underlying nanopattern with pitch of 600 nm was chosen as shown in Fig. 4.12a. The XRC-FWHM were determined with 100″ for the 0002 reflection and 450″ for the 10-12 reflection, which leads to a TDD of $2.4 \times 10^9\,cm^{-2}$ estimated by calculation. For preparing an AlN template with a similar TDD but a planar AlN/sapphire interface the HTA method described in chapter 3.3 was used. One of the outcomes of this method was that the TDD can be adjusted by the annealing parameters. A 900 nm thick AlN layer on planar sapphire was annealed at 1650 °C for 1 h. The XRC-FWHM of the HTA planar reference were determined with 70″ for the 0002 reflection and 520″ for the 10-12 reflection. This results in a TDD of $3.2 \times 10^9\,cm^{-2}$ estimated by calculation which is very similar to the AlN on top of the nanohole NPSS. The samples were cut into two pieces and overgrown with an LED heterostructure only up to the active region and a full LED heterostructure with a fully absorbing p-side. Details of the growth can be found in reference [135]. The LED heterostructures grown only up to the active region were measured by panchromatic CL at 80 K with acceleration voltage of 5 keV for the electron beam to determine the darkspot density (DSD). The measurement was performed by Carsten Netzel at FBH. The DSD is a direct measure of the non-radiative recombination centers, which correspond to the amount of TDs. So by determining the DSD the assumption can be checked if the TDD in the active region is comparable for both samples. The acquired images with the written DSDs are shown in Fig. 5.3. The DSD was counted to $2.2 \times 10^9\,cm^{-2}$ for the nanohole NPSS sample and $3.2 \times 10^9\,cm^{-2}$ for the HTA planar sample. Considering measurement errors, these densities can be considered as being comparable and are in good agreement with the TDDs of the underlying AlN templates. The LED heterostructures on top were grown pseudomorphically and thus, without significant generation of new TDs. For the full LED heterostructures, EL measurements including $L - I$ curves and emission spectra at 20 mA can be seen in Fig. 5.4. The emission power at 20 mA is 0.41 mW for the HTA planar sample and 0.4 mW for the nanohole NPSS sample. This approx. equal emission power leads to the conclusion that the nanopatterned AlN/sapphire interface in this case does not have a significant effect as predicted by the simulation results for the case of a fully absorbing p-side. The emission spectra show single peak emission for both samples at a wavelength of approx. 265 nm.

The same experiment was repeated for an LED heterostructure with transparent p-side. The AlN used this time was grown on top of the nanohole NPSS shown in Fig.

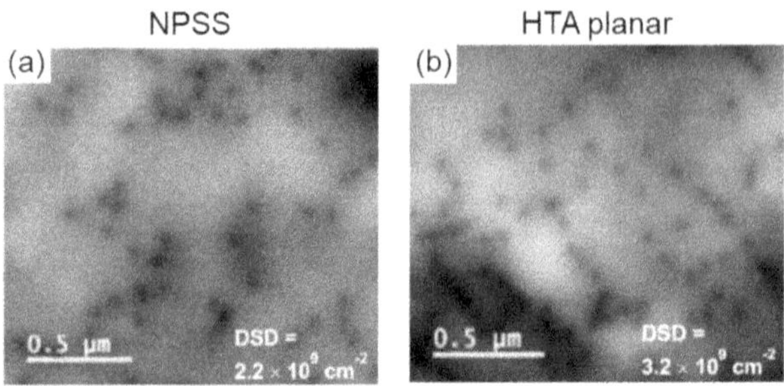

Figure 5.3: Panchromatic CL images of the surface of AlGaN MQW grown on top of the HTA planar and the NPSS template. The darkspot densities can be determined from the images by counting and are written down in the respective images. The HTA planar template was annealed at 1650 °C for 1 h. For the NPSS template a nanohole NPSS with pitch of 600 nm was used. From reference [135] (second authorship) with permission of John Wiley and Sons © 2020.

4.17b in chapter 4.3 with a pattern pitch of 1 µm. The XRC-FWHM were determined with 110″ for the 0002 reflection and 392″ for the 10-12 reflection resulting in a TDD of $1.8 \times 10^9\,\mathrm{cm}^{-2}$ estimated by calculation. As reference, a 900 nm thick AlN layer on planar sapphire was annealed at 1650 °C for 3 h. The XRC-FWHM of the HTA planar reference were determined with 52″ for the 0002 reflection and 380″ for the 10-12 reflection. The related TDD is $1.7 \times 10^9\,\mathrm{cm}^{-2}$ estimated by calculation. Hence, the TDDs of the NPSS and HTA planar template are comparable. Analogous to the previous experiment, panchromatic CL was conducted on the surface of subsequently grown AlGaN MQW on both samples. The related images are depicted in Fig. 5.5. The DSD was counted to $2.4 \times 10^9\,\mathrm{cm}^{-2}$ for the nanohole NPSS and $2 \times 10^9\,\mathrm{cm}^{-2}$ for the HTA planar template, which is also similar and in good agreement with the TDDs of the underlying AlN templates. As mentioned above, the full LED heterostructures were this time grown with a transparent p-side. Details of the growth process are described in chapter 2.1. The EL measurements of the full LED heterostructures grown on the HTA planar and nanohole NPSS template including $L-I$ curves and emission spectra at 20 mA are shown in Fig. 5.6. The LEDs with the transparent p-side in general show a much higher emission power compared to the LEDs of the previous experiments with absorbing p-side. This is due to the in general higher LEE [206]. However, the electrical efficiency will decrease due to the high Al-content of the p-side which leads to inferior p-contacts. This will reduce the wall plug efficiency. The emission power for

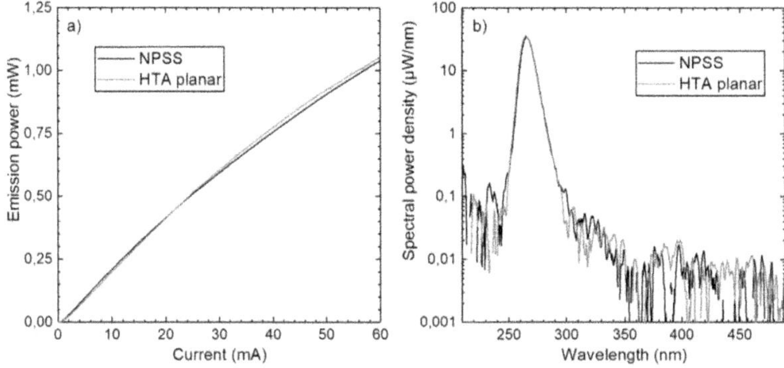

Figure 5.4: **(a)** $L-I$ curves and **(b)** emission spectra at a current of 20 mA for UVC LED heterostructures with absorbing p-side grown on top of the HTA planar and NPSS template. The HTA planar template was annealed at 1650 °C for 1 h. For the NPSS template a nanohole NPSS with pitch of 600 nm was used. From reference [135] (second authorship) with permission of John Wiley and Sons © 2020.

the nanohole NPSS sample is higher than for the HTA planar sample. At a current of 20 mA, the emission power is 2.16 mW for the nanohole NPSS sample and 1.61 mW for the HTA planar sample. This translates to an improvement of the emission power by approx. 34 % for the nanohole NPSS sample. The emission spectra at 20 mA (Fig. 5.6) show main luminescence at approx. 264 nm for both samples. For the nanohole NPSS sample a small side peak is observed for a wavelength of 327 nm. However, the intensity is more than 400-fold lower than the intensity of the main luminescence peak and therefore negligible. The increase in the emission power of 34 % at a current of 20 mA is ascribed to the implemented nanopatterned AlN/sapphire interface. Thus, the experimental results qualitatively are in agreement with the predictions of the wave optical simulations. The indium dots used as contact material have a reflectivity of approx. 87 % at a wavelength of 265 nm [207, 208]. Quantitatively, the simulation results predict a gain of the transmittance through the AlN/sapphire interface of 21 % from 57 % to 69 % for a nanohole pattern with a pitch of 1000 nm, a wavelength of 265 nm and isotropic light polarisation. Since in the experiment the emitted light is expected to be fully TE polarised [188], the gain of the transmittance will decrease as explained in chapter 4.1. Hence, the prediction of the simulation is in good agreement with the experimental results even quantitatively. However, quantitative predictions of the simulation results will still be limited, since scattering at the final sapphire/air

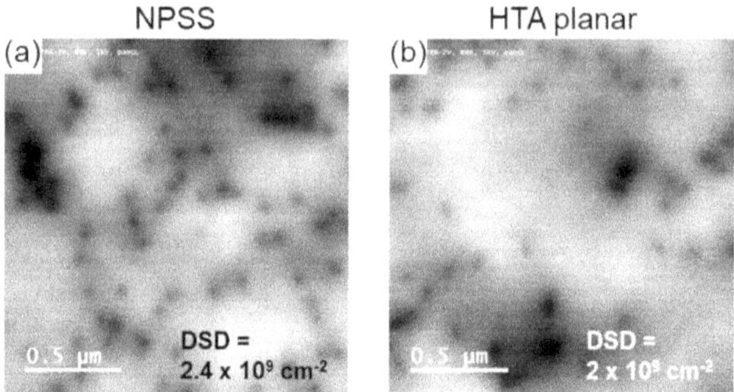

Figure 5.5: Panchromatic CL images of the surface of AlGaN MQW grown on top of the HTA planar and the NPSS template. The darkspot densities can be determined from the images by counting and are written down in the respective images. The HTA planar template was annealed at 1650 °C for 3 h. For the NPSS template a nanohole NPSS with pitch of 1000 nm was used.

interface will greatly influence the total LEE.

Summary In this chapter the AlN templates developed in this work were evaluated in terms of the emission power of UVC LED heterostructures with emission wavelength of 265 nm. Different comparisons were designed to estimate the improvement gained by the newly developed AlN templates:

1. The HTA planar template was compared with a planar template of 1.5 µm thickness without annealing. The emission power at 20 mA was improved more than 4-fold from 0.13 mW to 0.58 mW. This was attributed to the lower TDD for the HTA planar template of $8 \times 10^8\,\mathrm{cm}^{-2}$ in contrast to $3.9 \times 10^9\,\mathrm{cm}^{-2}$ for the planar template.

2. The comparison of the two AlN templates on top of the nanopillar NPSS with offcuts of 0.2° and 0.1° showed an increase in emission power for the LED heterostructure on the 0.1° offcut sample. At 20 mA the emission power increased more than 5-fold from 0.1 mW to 0.52 mW. The explanation for this was the pit formation during growth of the AlGaN heterostructure due to the step-bunched surface of the AlN for the 0.2° offcut sample. The pits did not show luminescence at the targeted wavelength of 265 nm but unwanted luminescence at 326 nm due to their Ga-rich filling.

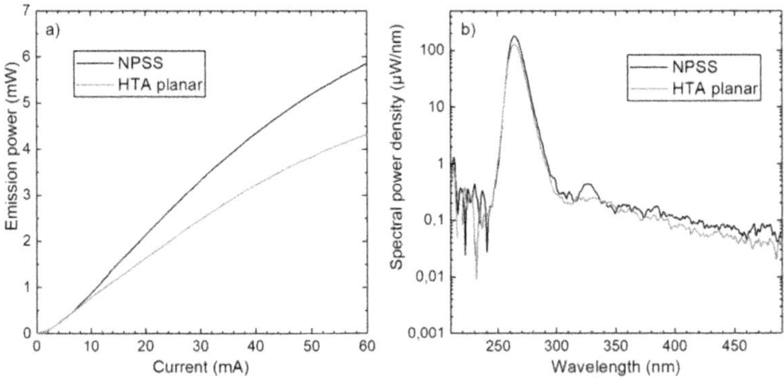

Figure 5.6: (a) $L-I$ curves and **(b)** emission spectra at a current of 20 mA for UVC LED heterostructures with transparent p-side grown on top of the HTA planar and NPSS template. The HTA planar template was annealed at 1650 °C for 3 h. For the NPSS template a nanohole NPSS with pitch of 1000 nm was used.

3. The AlN template on top of the nanohole NPSS was compared with an HTA planar template. The TDD in the active regions of the LEDs was checked to be similar by determining the DSDs. For a fully absorbing p-side, no gain in emission power was observed, which is in agreement with the simulation results. For a transparent p-side combined with 87 % reflectivity of the indium contact, the emission power at 20 mA could be increased by 34 % from 1.61 mW to 2.16 mW for the nanohole NPSS sample. This result is again in good agreement with the simulation results and shows that the light extraction can be improved by an implemented nanopatterned AlN/sapphire interface.

In all, UV LED emission was demonstrated on all newly developed templates, which proves the suitability to use them as UV LED base layers. An outlook what template would overall be the best one can be given as follows. It was shown that it is crucial to have a low TDD to increase the IQE on one hand. But on the other hand the nanopatterned interface enables an increase of the LEE in case of an UV LED with transparent p-side and reflective contact material. Unfortunately, the TDD reached by overgrowth of the NPSS was not as low as expected so that further work has to be done regarding this issue. In the end, a nanopatterned interface should be combined with a low TDD of the AlN. A good opportunity for this would be to use a sputtered AlN layer as a starting layer on top of the NPSS, since sputtered layers can show a more efficient TDD reduction during HTA [117]. A sputtered starting layer would even make the whole process easier, since the sensitive nucleation on the sapphire inside the MOVPE can be skipped. Recently, there has been a report in literature demonstrating the realisation of a sputtered AlN layer on top of the NPSS and subsequent MOVPE overgrowth with a final TDD as low as $6 \times 10^8\,\mathrm{cm}^{-2}$ [120]. The sputtered AlN starting layer on nanohole NPSS, subsequent HTA and overgrowth has the potential to combine the benefits of all different methods: A low TDD, nanopatterned interface and thin final layer thickness resulting in small wafer bow. Additionally, the sputtering makes the process even more easy and stable in terms of reproducibility. At last, in terms of process costs, commercial availability of the substrate is desirable. This is realised for the nanohole NPSS at a very reasonable price making the technology very cost effective.

Conclusion

In this work, the novel techniques of high temperature annealing and nanopatterned sapphire substrates for preparation of AlN base layers have been investigated. The aim was to improve the performance of subsequently grown UV LED heterostructures by reducing the threading dislocation density in the AlN base layer to increase the internal quantum efficiency and by implementing a nanopatterned AlN/sapphire interface to increase the light extraction efficiency.

By HTA of MOVPE grown AlN/sapphire layers in face-to-face configuration, the TDD could be reduced by more than one order of magnitude down to $6 \times 10^8\,\text{cm}^{-2}$. This was achieved by annealing at 1700 °C for 3 h of 900 nm thick AlN, which was grown with a two-step growth process to induce roughening and subsequent smoothing of the surface to suppress the formation of tensile strain. If too much tensile strain is incorporated in the AlN layers, they will crack during subsequent HTA. After annealing, the AlN layers showed a change of the strain to more compressive values. This has to be considered for AlGaN growth on top. Although after annealing a high amount of oxygen was found associated with diffusion from the sapphire substrate into the AlN layer resulting in the formation of AlON, no effect on the transmission of the AlN template in the UV spectral range was observed.

It was found that the usage of NPSS has the potential to increase the light extraction efficiency of UV LED heterostructures, if an at least partly reflecting p-side is realised on top. For a nanohole pattern with pitch of 1000 nm, hole diameter of 500 nm, hole depth of 250 nm and an assumed effective reflectivity of the p-side of 0.8, wave optical simulation showed an increase in the transmittance through the AlN/sapphire interface by 77 %. This was explained by increased scattering at the nanopatterned AlN/interface softening the issue of total internal reflection. Successful AlN overgrowth of a nanohole and nanopillar pattern could be realised obtaining an atomically smooth surface. By reducing the offcut of the nanopillar-patterned sapphire, the formation of high step bunches, which are detrimental for further AlGaN growth, could be prevented. For growth on sapphire with the nanohole pattern, the continuous surface at growth start led to cracking of the AlN layer prior to coalescence. An intermediate HTA step was introduced for changing the strain to more compressive values. Subsequent AlN overgrowth succeeded in full coalescence without any layer cracking. The TDD showed similar values for AlN on the nanopillars and nanoholes of approx. $1 \times 10^9\,\mathrm{cm}^{-2}$ which are due to the generation of new TDs at coalescence. The AlN on top of the nanoholes coalesced considerably faster than on the nanopillars so that the AlN layer thickness could be reduced from 7 µm for growth on the nanopillars down to 2 µm for growth on the nanoholes.

The suitability of the developed AlN templates was checked by determining the performance of UVC LEDs emitting at 265 nm grown on top. Electroluminescence was achieved for all AlN templates. Specifically designed experiments showed considerable improvement by using the newly developed AlN templates. Furthermore, by comparing the performance of UVC LEDs with transparent p-side on two AlN templates with the same material quality but with and without a nanopatterned AlN/sapphire interface, the experimental proof was given for the prediction of the wave optical simulations regarding the increased light extraction efficiency for the nanopatterned AlN/sapphire interface.

Outlook The present work shows that both a low TDD and a nanopatterned AlN/sapphire interface can improve the performance of UV LEDs grown on top. The next step now is to further reduce the TDD of the AlN templates on the NPSS, since with a TDD of approx. $1 \times 10^9\,\mathrm{cm}^{-2}$ there is still room for improvement in internal quantum efficiency. A promising roadmap for that is the implementation of a sputtered starting layer rather than an MOVPE grown starting layer for overgrowth of nanohole-patterned sapphire. HTA of sputtered AlN layers in general enables lower TDDs compared with annealed MOVPE grown layers. Very recent work by Wang et al. shows that by proficient combination of AlN sputtering and HTA, TDDs in the mid $10^7\,\mathrm{cm}^{-2}$ are realistic [125]. Additionally, with the use of sputtering instead of MOVPE growth the whole process becomes less complex and costly. The AlN sputtering and HTA combined with the commercial supply of NPSS enables a very cost-effective technology with the potential of industrial mass production. This can be a big milestone on the roadmap of establishing UV LEDs in the near future.

APPENDIX A

Appendix

Bibliography

[1] R. D. Dupuis, M. R. Krames, *History, Development, and Applications of High-Brightness Visible Light-Emitting Diodes*, Journal of Lightwave Technology **26**, 1154 (2008).

[2] J. Cho, J. H. Park, J. K. Kim, E. F. Schubert, *White light-emitting diodes: History, progress, and future*, Laser & Photonics Reviews **11**, 1600147 (2017).

[3] J. Heber, *Nobel Prize 2014: Akasaki, Amano & Nakamura*, Nature Physics **10**, 791 (2014).

[4] I. Akasaki, H. Amano, K. Hiramatsu, N. Sawaki, *High efficiency blue LED utilizing GaN film with AlN buffer layer grown by MOVPE*, Institute of Physics Conference Series **91**, 633 (1988).

[5] S. Nakamura, T. Mukai, M. Senoh, *Candela-class high-brightness InGaN/AlGaN double-heterostructure blue-light-emitting diodes*, Applied Physics Letters **64**, 1687 (1994).

[6] M. Kneissl, J. Rass, Eds., *III-Nitride Ultraviolet Emitters*, Springer International Publishing, 2016.

[7] F. Mehnke, M. Guttmann, J. Enslin, C. Kuhn, C. Reich, J. Jordan, S. Kapanke, A. Knauer, M. Lapeyrade, U. Zeimer, H. Kruger, M. Rabe, S. Einfeldt, T. Wernicke, H. Ewald, M. Weyers, M. Kneissl, *Gas Sensing of Nitrogen Oxide Utilizing*

Spectrally Pure Deep UV LEDs, IEEE Journal of Selected Topics in Quantum Electronics **23**, 29 (2017).

[8] J. Mellqvist, A. Rosén, *DOAS for flue gas monitoring—I. Temperature effects in the U.V./visible absorption spectra of NO, NO2, SO2 and NH3*, Journal of Quantitative Spectroscopy and Radiative Transfer **56**, 187 (1996).

[9] J. Hodgkinson, R. P. Tatam, *Optical gas sensing: a review*, Measurement Science and Technology **24**, 012004 (2012).

[10] *Applications for UV LEDs*, Advanced UV for Life, https://www.advanced-uv.de/en/uv-basics/uv-radiation (28.09.2020).

[11] B. Song, H. Sun, C.-K. Kao, T. D. Moustakas, *Growth mode of nitride semiconductors on nano-patterned sapphire substrates by molecular beam epitaxy*, physica status solidi (c) **13**, 195 (2016).

[12] S. Vilhunen, H. Särkkä, M. Sillanpää, *Ultraviolet light-emitting diodes in water disinfection*, Environmental Science and Pollution Research **16**, 439 (2009).

[13] M. Würtele, T. Kolbe, M. Lipsz, A. Külberg, M. Weyers, M. Kneissl, M. Jekel, *Application of GaN-based ultraviolet-C light emitting diodes – UV LEDs – for water disinfection*, Water Research **45**, 1481 (2011).

[14] G. Y. Lui, D. Roser, R. Corkish, N. Ashbolt, P. Jagals, R. Stuetz, *Photovoltaic powered ultraviolet and visible light-emitting diodes for sustainable point-of-use disinfection of drinking waters*, Science of The Total Environment **493**, 185 (2014).

[15] J. C. Chang, S. F. Ossoff, D. C. Lobe, M. H. Dorfman, C. M. Dumais, R. G. Qualls, J. D. Johnson, *UV inactivation of pathogenic and indicator microorganisms.*, Applied and Environmental Microbiology **49**, 1361 (1985).

[16] Y. W. Lee, H. D. Yoon, J.-H. Park, U.-C. Ryu, *Application of 265-nm UVC LED Lighting to Sterilization of Typical Gram Negative and Positive Bacteria*, Journal of the Korean Physical Society **72**, 1174 (2018).

[17] M. E. Darnell, K. Subbarao, S. M. Feinstone, D. R. Taylor, *Inactivation of the coronavirus that induces severe acute respiratory syndrome, SARS-CoV*, Journal of Virological Methods **121**, 85 (2004).

[18] S. M. Duan, X. S. Zhao, R. F. Wen, J. J. Huang, G. H. Pi, S. X. Zhang, J. Han, S. L. Bi, L. Ruan, X. P. Dong, *Stability of SARS coronavirus in human specimens and environment and its sensitivity to heating and UV irradiation*, Biomedical and Environmental Sciences : BES **16**, 246 (2003).

[19] Y. Tsunetsugu-Yokota, *Large-Scale Preparation of UV-Inactivated SARS Coronavirus Virions for Vaccine Antigen*, in *Methods in Molecular Biology*, pages 119–126, Humana Press, 2008.

[20] H. Inagaki, A. Saito, H. Sugiyama, T. Okabayashi, S. Fujimoto, *Rapid inactivation of SARS-CoV-2 with deep-UV LED irradiation*, Emerging Microbes & Infections **9**, 1744 (2020).

[21] M. Schreiner, J. Martínez-Abaigar, J. Glaab, M. Jansen, *UV-B Induced Secondary Plant Metabolites*, Optik & Photonik **9**, 34 (2014).

[22] F. Bantis, S. Smirnakou, T. Ouzounis, A. Koukounaras, N. Ntagkas, K. Radoglou, *Current status and recent achievements in the field of horticulture with the use of light-emitting diodes (LEDs)*, Scientia Horticulturae **235**, 437 (2018).

[23] P. E. Hockberger, *A History of Ultraviolet Photobiology for Humans, Animals and Microorganisms*, Photochemistry and Photobiology **76**, 561 (2002).

[24] J. Philip J. Hargis, T. J. Sobering, G. C. Tisone, J. S. Wagner, S. A. Young, R. J. Radloff, *Ultraviolet fluorescence identification of protein, DNA, and bacteria*, Optical Instrumentation for Gas Emissions Monitoring and Atmospheric Measurements **2366**, 147 (1995).

[25] K. C. Anyaogu, A. A. Ermoshkin, D. C. Neckers, A. Mejiritski, O. Grinevich, A. V. Fedorov, *Performance of the light emitting diodes versus conventional light sources in the UV light cured formulations*, Journal of Applied Polymer Science **105**, 803 (2007).

[26] A. Endruweit, M. S. Johnson, A. C. Long, *Curing of composite components by ultraviolet radiation: A review*, Polymer Composites **27**, 119 (2006).

[27] C. Decker, *The use of UV irradiation in polymerization*, Polymer International **45**, 133 (1998).

[28] M. Brozel, *Gallium Arsenide*, in *Springer Handbook of Electronic and Photonic Materials*, pages 499–536, Springer US, 2006.

[29] H. Amano, *Progress and Prospect of the Growth of Wide-Band-Gap Group III Nitrides: Development of the Growth Method for Single-Crystal Bulk GaN*, Japanese Journal of Applied Physics **52**, 050001 (2013).

[30] W.-H. Chen, Z.-Y. Qin, X.-Y. Tian, X.-H. Zhong, Z.-H. Sun, B.-K. Li, R.-S. Zheng, Y. Guo, H.-L. Wu, *The Physical Vapor Transport Method for Bulk AlN Crystal Growth*, Molecules **24**, 1562 (2019).

[31] R. T. Bondokov, S. G. Mueller, K. E. Morgan, G. A. Slack, S. Schujman, M. C. Wood, J. A. Smart, L. J. Schowalter, *Large-area AlN substrates for electronic applications: An industrial perspective*, Journal of Crystal Growth **310**, 4020 (2008).

[32] R. Dalmau, B. Moody, J. Xie, R. Collazo, Z. Sitar, *Characterization of dislocation arrays in AlN single crystals grown by PVT*, physica status solidi (a) **208**, 1545 (2011).

[33] C. Hartmann, J. Wollweber, S. Sintonen, A. Dittmar, L. Kirste, S. Kollowa, K. Irmscher, M. Bickermann, *Preparation of deep UV transparent AlN substrates with high structural perfection for optoelectronic devices*, CrystEngComm **18**, 3488 (2016).

[34] C. Hartmann, A. Dittmar, J. Wollweber, M. Bickermann, *Bulk AlN growth by physical vapour transport*, Semiconductor Science and Technology **29**, 084002 (2014).

[35] M. Bickermann, *Growth and Properties of Bulk AlN Substrates*, in *III-Nitride Ultraviolet Emitters*, pages 27–46, Springer International Publishing, 2015.

[36] S. Nakamura, M. R. Krames, *History of Gallium–Nitride-Based Light-Emitting Diodes for Illumination*, Proceedings of the IEEE **101**, 2211 (2013).

[37] G. Li, W. Wang, W. Yang, Y. Lin, H. Wang, Z. Lin, S. Zhou, *GaN-based light-emitting diodes on various substrates: a critical review*, Reports on Progress in Physics **79**, 056501 (2016).

[38] D. Nilsson, E. Janzén, A. Kakanakova-Georgieva, *Lattice parameters of AlN bulk, homoepitaxial and heteroepitaxial material*, Journal of Physics D: Applied Physics **49**, 175108 (2016).

[39] M. Feneberg, R. A. R. Leute, B. Neuschl, K. Thonke, M. Bickermann, *High-excitation and high-resolution photoluminescence spectra of bulk AlN*, Physical Review B **82**, 075208 (2010).

[40] E. S. Hellman, *The Polarity of GaN: a Critical Review*, MRS Internet Journal of Nitride Semiconductor Research **3**, E11 (1998).

[41] N. Stolyarchuk, T. Markurt, A. Courville, K. March, J. Zúñiga-Pérez, P. Vennéguès, M. Albrecht, *Intentional polarity conversion of AlN epitaxial layers by oxygen*, Scientific Reports **8**, 14111 (2018).

[42] V. Küller, *Versetzungsreduzierte AlN- und AlGaN-Schichten als Basis für UV LEDs*, Dissertation, Technische Universität Berlin (2014).

[43] W. M. Yim, R. J. Paff, *Thermal expansion of AlN, sapphire, and silicon*, Journal of Applied Physics **45**, 1456 (1974).

[44] B. B. Kosicki, D. Kahng, *Preparation and Structural Properties of GaN Thin Films*, Journal of Vacuum Science and Technology **6**, 593 (1969).

[45] T. Sasaki, S. Zembutsu, *Substrate-orientation dependence of GaN single-crystal films grown by metalorganic vapor-phase epitaxy*, Journal of Applied Physics **61**, 2533 (1987).

[46] H. M. Manasevit, F. M. Erdmann, W. I. Simpson, *The Use of Metalorganics in the Preparation of Semiconductor Materials*, Journal of The Electrochemical Society **118**, 1864 (1971).

[47] P. Kung, C. J. Sun, A. Saxler, H. Ohsato, M. Razeghi, *Crystallography of epitaxial growth of wurtzite-type thin films on sapphire substrates*, Journal of Applied Physics **75**, 4515 (1994).

[48] C. J. Sun, P. Kung, A. Saxler, H. Ohsato, K. Haritos, M. Razeghi, *A crystallographic model of (00·1) aluminum nitride epitaxial thin film growth on (00·1) sapphire substrate*, Journal of Applied Physics **75**, 3964 (1994).

[49] A. E. Romanov, G. E. Beltz, P. Cantu, F. Wu, S. Keller, S. P. DenBaars, J. S. Speck, *Cracking of III-nitride layers with strain gradients*, Applied Physics Letters **89**, 161922 (2006).

[50] D. Holec, P. Costa, M. Kappers, C. Humphreys, *Critical thickness calculations for InGaN/GaN*, Journal of Crystal Growth **303**, 314 (2007).

[51] D. Holec, Y. Zhang, D. V. S. Rao, M. J. Kappers, C. McAleese, C. J. Humphreys, *Equilibrium critical thickness for misfit dislocations in III-nitrides*, Journal of Applied Physics **104**, 123514 (2008).

[52] X. J. Ning, F. R. Chien, P. Pirouz, J. W. Yang, M. A. Khan, *Growth defects in GaN films on sapphire: The probable origin of threading dislocations*, Journal of Materials Research **11**, 580 (1996).

[53] Z.-M. Chen, Z.-Y. Zheng, Y.-D. Chen, H.-L. Wu, C.-S. Tong, G. Wang, Z.-S. Wu, H. Jiang, *Threading edge dislocation arrays in epitaxial GaN: Formation, model and thermodynamics*, Journal of Crystal Growth **387**, 48 (2014).

[54] X. Wu, P. Fini, E. Tarsa, B. Heying, S. Keller, U. Mishra, S. DenBaars, J. Speck, *Dislocation generation in GaN heteroepitaxy*, Journal of Crystal Growth **189-190**, 231 (1998).

[55] O. Reentilä, F. Brunner, A. Knauer, A. Mogilatenko, W. Neumann, H. Protzmann, M. Heuken, M. Kneissl, M. Weyers, G. Tränkle, *Effect of the AIN nucleation layer growth on AlN material quality*, Journal of Crystal Growth **310**, 4932 (2008).

[56] U. W. Pohl, *Epitaxy of Semiconductors*, Springer Berlin Heidelberg, 2013.

[57] J. P. Hirth, J. Lothe, T. Mura, *Theory of Dislocations (2nd ed.)*, Journal of Applied Mechanics **50**, 476 (1983).

[58] F. Scholz, *Compound Semiconductors*, Jenny Stanford Publishing, 2017.

[59] E. Rosencher, B. Vinter, *Light emitting diodes and laser diodes*, in *Optoelectronics*, pages 613–659, Cambridge University Press, 2002.

[60] S. Sze, K. K. Ng, *Physics of Semiconductor Devices*, John Wiley & Sons, Inc., 2006.

[61] S. L. Chuang, N. Peyghambarian, S. Koch, *Physics of Optoelectronic Devices*, Physics Today **49**, 62 (1996).

[62] M. Kneissl, T.-Y. Seong, J. Han, H. Amano, *The emergence and prospects of deep-ultraviolet light-emitting diode technologies*, Nature Photonics **13**, 233 (2019).

[63] M. Kneissl, T. Kolbe, C. Chua, V. Kueller, N. Lobo, J. Stellmach, A. Knauer, H. Rodriguez, S. Einfeldt, Z. Yang, N. M. Johnson, M. Weyers, *Advances in group III-nitride-based deep UV light-emitting diode technology*, Semiconductor Science and Technology **26**, 014036 (2010).

[64] J. Mickevičius, G. Tamulaitis, M. Shur, M. Shatalov, J. Yang, R. Gaska, *Internal quantum efficiency in AlGaN with strong carrier localization*, Applied Physics Letters **101**, 211902 (2012).

[65] K. Ban, J. ichi Yamamoto, K. Takeda, K. Ide, M. Iwaya, T. Takeuchi, S. Kamiyama, I. Akasaki, H. Amano, *Internal Quantum Efficiency of Whole-Composition-Range AlGaN Multiquantum Wells*, Applied Physics Express **4**, 052101 (2011).

[66] H. Hirayama, S. Fujikawa, N. Noguchi, J. Norimatsu, T. Takano, K. Tsubaki, N. Kamata, *222-282 nm AlGaN and InAlGaN-based deep-UV LEDs fabricated on high-quality AlN on sapphire*, physica status solidi (a) **206**, 1176 (2009).

[67] M. Shatalov, W. Sun, A. Lunev, X. Hu, A. Dobrinsky, Y. Bilenko, J. Yang, M. Shur, R. Gaska, C. Moe, G. Garrett, M. Wraback, *AlGaN Deep-Ultraviolet Light-Emitting Diodes with External Quantum Efficiency above 10%*, Applied Physics Express **5**, 082101 (2012).

[68] T.-Y. Wang, C.-T. Tasi, C.-F. Lin, D.-S. Wuu, *85% internal quantum efficiency of 280-nm AlGaN multiple quantum wells by defect engineering*, Scientific Reports **7** (2017).

[69] S. Y. Karpov, Y. N. Makarov, *Dislocation effect on light emission efficiency in gallium nitride*, Applied Physics Letters **81**, 4721 (2002).

[70] I. Bryan, Z. Bryan, S. Washiyama, P. Reddy, B. Gaddy, B. Sarkar, M. H. Breckenridge, Q. Guo, M. Bobea, J. Tweedie, S. Mita, D. Irving, R. Collazo, Z. Sitar, *Doping and compensation in Al-rich AlGaN grown on single crystal AlN and sapphire by MOCVD*, Applied Physics Letters **112**, 062102 (2018).

[71] Y. Nagasawa, A. Hirano, *A Review of AlGaN-Based Deep-Ultraviolet Light-Emitting Diodes on Sapphire*, Applied Sciences **8**, 1264 (2018).

[72] H. P. Maruska, J. J. Tietjen, *The prparation and properties of vapor-deposited single-crystal-line GaN*, Applied Physics Letters **15**, 327 (1969).

[73] S. Hagedorn, S. Walde, A. Knauer, N. Susilo, D. Pacak, L. Cancellara, C. Netzel, A. Mogilatenko, C. Hartmann, T. Wernicke, M. Kneissl, M. Weyers, *Status and Prospects of AlN Templates on Sapphire for Ultraviolet Light-Emitting Diodes*, physica status solidi (a) **217**, 1901022 (2020).

[74] A. C. Jones, S. A. Rushworth, D. J. Houlton, J. S. Roberts, V. Roberts, C. R. Whitehouse, G. W. Critchlow, *Deposition of aluminum nitride thin films by MOCVD from the trimethylaluminum-ammonia adduct*, Chemical Vapor Deposition **2**, 5 (1996).

[75] Y. Kumagai, T. Yamane, A. Koukitu, *Growth of thick AlN layers by hydride vapor-phase epitaxy*, Journal of Crystal Growth **281**, 62 (2005).

[76] X.-H. Liu, J.-C. Zhang, J. Huang, M.-M. Yang, X.-J. Su, B.-B. Ye, J.-F. Wang, J.-P. Zhang, K. Xu, *Influence of growth temperature on intrinsic stress distribution in aluminum nitride grown by hydride vapor phase epitaxy*, Materials Express **6**, 367 (2016).

[77] X.-Q. Shen, M. Shimizu, H. Okumura, *Impact of Vicinal Sapphire (0001) Substrates on the High-Quality AlN Films by Plasma-Assisted Molecular Beam Epitaxy*, Japanese Journal of Applied Physics **42**, L1293 (2003).

[78] B. Kuhn, F. Scholz, *An Oxygen Doped Nucleation Layer for the Growth of High Optical Quality GaN on Sapphire*, physica status solidi (a) **188**, 629 (2001).

[79] J. Bläsing, A. Krost, J. Hertkorn, F. Scholz, L. Kirste, A. Chuvilin, U. Kaiser, *Oxygen induced strain field homogenization in AlN nucleation layers and its impact on GaN grown by metal organic vapor phase epitaxy on sapphire: An x-ray diffraction study*, Journal of Applied Physics **105**, 033504 (2009).

[80] D. D. Koleske, J. J. Figiel, D. L. Alliman, B. P. Gunning, J. M. Kempisty, J. R. Creighton, A. Mishima, K. Ikenaga, *Metalorganic vapor phase epitaxy of AlN on sapphire with low etch pit density*, Applied Physics Letters **110**, 232102 (2017).

[81] S. Mohn, N. Stolyarchuk, T. Markurt, R. Kirste, M. P. Hoffmann, R. Collazo, A. Courville, R. D. Felice, Z. Sitar, P. Vennéguès, M. Albrecht, *Polarity Control in Group-III Nitrides beyond Pragmatism*, Physical Review Applied **5**, 054004 (2016).

[82] M. Funato, M. Shibaoka, Y. Kawakami, *Heteroepitaxy mechanisms of AlN on nitridated c- and a-plane sapphire substrates*, Journal of Applied Physics **121**, 085304 (2017).

[83] H. Sun, F. Wu, Y. J. Park, T. M. A. tahtamouni, K.-H. Li, N. Alfaraj, T. Detchprohm, R. D. Dupuis, X. Li, *Influence of TMAl preflow on AlN epitaxy on sapphire*, Applied Physics Letters **110**, 192106 (2017).

[84] N. Okada, N. Kato, S. Sato, T. Sumii, T. Nagai, N. Fujimoto, M. Imura, K. Balakrishnan, M. Iwaya, S. Kamiyama, H. Amano, I. Akasaki, H. Maruyama, T. Takagi, T. Noro, A. Bandoh, *Growth of high-quality and crack free AlN layers on sapphire substrate by multi-growth mode modification*, Journal of Crystal Growth **298**, 349 (2007).

[85] H. Hirayama, T. Yatabe, N. Noguchi, T. Ohashi, N. Kamata, *231–261nm AlGaN deep-ultraviolet light-emitting diodes fabricated on AlN multilayer buffers grown by ammonia pulse-flow method on sapphire*, Applied Physics Letters **91**, 071901 (2007).

[86] J. Yan, J. Wang, Y. Zhang, P. Cong, L. Sun, Y. Tian, C. Zhao, J. Li, *AlGaN-based deep-ultraviolet light-emitting diodes grown on High-quality AlN template using MOVPE*, Journal of Crystal Growth **414**, 254 (2015).

[87] T. M. A. tahtamouni, J. Y. Lin, H. X. Jiang, *Effects of double layer AlN buffer layers on properties of Si-doped AlxGa1-xN for improved performance of deep ultraviolet light emitting diodes*, Journal of Applied Physics **113**, 123501 (2013).

[88] K. Nakano, M. Imura, G. Narita, T. Kitano, Y. Hirose, N. Fujimoto, N. Okada, T. Kawashima, K. Iida, K. Balakrishnan, M. Tsuda, M. Iwaya, S. Kamiyama, H. Amano, I. Akasaki, *Epitaxial lateral overgrowth of AlN layers on patterned sapphire substrates*, physica status solidi (a) **203**, 1632 (2006).

[89] M. Imura, K. Nakano, T. Kitano, N. Fujimoto, G. Narita, N. Okada, K. Balakrishnan, M. Iwaya, S. Kamiyama, H. Amano, I. Akasaki, K. Shimono, T. Noro, T. Takagi, A. Bandoh, *Microstructure of epitaxial lateral overgrown AlN on trench-patterned AlN template by high-temperature metal-organic vapor phase epitaxy*, Applied Physics Letters **89**, 221901 (2006).

[90] H. Hirayama, S. Fujikawa, J. Norimatsu, T. Takano, K. Tsubaki, N. Kamata, *Fabrication of a low threading dislocation density ELO-AlN template for application to deep-UV LEDs*, physica status solidi (c) **6**, S356 (2009).

[91] V. Kueller, A. Knauer, U. Zeimer, H. Rodriguez, A. Mogilatenko, M. Kneissl, M. Weyers, *(Al,Ga)N overgrowth over AlN ridges oriented in [1120] and [1100] direction*, physica status solidi (c) **8**, 2022 (2011).

[92] H.-Y. Wang, Z.-T. Lin, J.-L. Han, L.-Y. Zhong, G.-Q. Li, *Design of patterned sapphire substrates for GaN-based light-emitting diodes*, Chinese Physics B **24**, 067103 (2015).

[93] L.-X. Zhao, Z.-G. Yu, B. Sun, S.-C. Zhu, P.-B. An, C. Yang, L. Liu, J.-X. Wang, J.-M. Li, *Progress and prospects of GaN-based LEDs using nanostructures*, Chinese Physics B **24**, 068506 (2015).

[94] P. Dong, J. Yan, J. Wang, Y. Zhang, C. Geng, T. Wei, P. Cong, Y. Zhang, J. Zeng, Y. Tian, L. Sun, Q. Yan, J. Li, S. Fan, Z. Qin, *282-nm AlGaN-based deep ultraviolet light-emitting diodes with improved performance on nano-patterned sapphire substrates*, Applied Physics Letters **102**, 241113 (2013).

[95] P. Dong, J. Yan, Y. Zhang, J. Wang, J. Zeng, C. Geng, P. Cong, L. Sun, T. Wei, L. Zhao, Q. Yan, C. He, Z. Qin, J. Li, *AlGaN-based deep ultraviolet light-emitting diodes grown on nano-patterned sapphire substrates with significant improvement in internal quantum efficiency*, Journal of Crystal Growth **395**, 9 (2014).

[96] B. T. Tran, H. Hirayama, N. Maeda, M. Jo, S. Toyoda, N. Kamata, *Direct Growth and Controlled Coalescence of Thick AlN Template on Micro-circle Patterned Si Substrate*, Scientific Reports **5**, 14734 (2015).

[97] S. Hagedorn, A. Knauer, A. Mogilatenko, E. Richter, M. Weyers, *AlN growth on nano-patterned sapphire: A route for cost efficient pseudo substrates for deep UV LEDs*, physica status solidi (a) **213**, 3178 (2016).

[98] L. Zhang, F. Xu, J. Wang, C. He, W. Guo, M. Wang, B. Sheng, L. Lu, Z. Qin, X. Wang, B. Shen, *High-quality AlN epitaxy on nano-patterned sapphire substrates prepared by nano-imprint lithography*, Scientific Reports **6**, 35934 (2016).

[99] D. Lee, J. W. Lee, J. Jang, I.-S. Shin, L. Jin, J. H. Park, J. Kim, J. Lee, H.-S. Noh, Y.-I. Kim, Y. Park, G.-D. Lee, Y. Park, J. K. Kim, E. Yoon, *Improved*

performance of AlGaN-based deep ultraviolet light-emitting diodes with nano-patterned AlN/sapphire substrates, Applied Physics Letters **110**, 191103 (2017).

[100] N. Xie, F. Xu, N. Zhang, J. Lang, J. Wang, M. Wang, Y. Sun, B. Liu, W. Ge, Z. Qin, X. Kang, X. Yang, X. Wang, B. Shen, *Period size effect induced crystalline quality improvement of AlN on a nano-patterned sapphire substrate*, Japanese Journal of Applied Physics **58**, 100912 (2019).

[101] N. Xie, F. Xu, J. Wang, Y. Sun, B. Liu, N. Zhang, J. Lang, X. Fang, W. Ge, Z. Qin, X. Kang, X. Yang, X. Wang, B. Shen, *Stress evolution in AlN growth on nano-patterned sapphire substrates*, Applied Physics Express **13**, 015504 (2019).

[102] H. Miyake, G. Nishio, S. Suzuki, K. Hiramatsu, H. Fukuyama, J. Kaur, N. Kuwano, *Annealing of an AlN buffer layer in N2–CO for growth of a high-quality AlN film on sapphire*, Applied Physics Express **9**, 025501 (2016).

[103] H. Fukuyama, H. Miyake, G. Nishio, S. Suzuki, K. Hiramatsu, *Impact of high-temperature annealing of AlN layer on sapphire and its thermodynamic principle*, Japanese Journal of Applied Physics **55**, 05FL02 (2016).

[104] J. Kaur, N. Kuwano, K. R. Jamaludin, M. Mitsuhara, H. Saito, S. Hata, S. Suzuki, H. Miyake, K. Hiramatsu, H. Fukuyama, *Electron microscopy analysis of microstructure of postannealed aluminum nitride template*, Applied Physics Express **9**, 065502 (2016).

[105] H. Miyake, C.-H. Lin, K. Tokoro, K. Hiramatsu, *Preparation of high-quality AlN on sapphire by high-temperature face-to-face annealing*, Journal of Crystal Growth **456**, 155 (2016).

[106] C.-Y. Huang, P.-Y. Wu, K.-S. Chang, Y.-H. Lin, W.-C. Peng, Y.-Y. Chang, J.-P. Li, H.-W. Yen, Y. S. Wu, H. Miyake, H.-C. Kuo, *High-quality and highly-transparent AlN template on annealed sputter-deposited AlN buffer layer for deep ultra-violet light-emitting diodes*, AIP Advances **7**, 055110 (2017).

[107] N. Susilo, S. Hagedorn, D. Jaeger, H. Miyake, U. Zeimer, C. Reich, B. Neuschulz, L. Sulmoni, M. Guttmann, F. Mehnke, C. Kuhn, T. Wernicke, M. Weyers, M. Kneissl, *AlGaN-based deep UV LEDs grown on sputtered and high temperature annealed AlN/sapphire*, Applied Physics Letters **112**, 041110 (2018).

[108] R. Yoshizawa, H. Miyake, K. Hiramatsu, *Effect of thermal annealing on AlN films grown on sputtered AlN templates by metalorganic vapor phase epitaxy*, Japanese Journal of Applied Physics **57**, 01AD05 (2017).

[109] J. Hakamata, Y. Kawase, L. Dong, S. Iwayama, M. Iwaya, T. Takeuchi, S. Kamiyama, H. Miyake, I. Akasaki, *Growth of High-Quality AlN and AlGaN Films on Sputtered AlN/Sapphire Templates via High-Temperature Annealing*, physica status solidi (b) **255**, 1700506 (2018).

[110] S. Xiao, R. Suzuki, H. Miyake, S. Harada, T. Ujihara, *Improvement mechanism of sputtered AlN films by high-temperature annealing*, Journal of Crystal Growth **502**, 41 (2018).

[111] T. Akiyma, M. Uchino, K. Nakamura, T. Ito, S. Xiao, H. Miyake, *Structural analysis of polarity inversion boundary in sputtered AlN films annealed under high temperatures*, Japanese Journal of Applied Physics **58**, SCCB30 (2019).

[112] Y. Hayashi, K. Tanigawa, K. Uesugi, K. Shojiki, H. Miyake, *Curvature-controllable and crack-free AlN/sapphire templates fabricated by sputtering and high-temperature annealing*, Journal of Crystal Growth **512**, 131 (2019).

[113] K. Nagamatsu, X. Liu, K. Uesugi, H. Miyake, *Improved emission intensity of UVC-LEDs from using strain relaxation layer on sputter-annealed AlN*, Japanese Journal of Applied Physics **58**, SCCC07 (2019).

[114] K. Shojiki, R. Ishii, K. Uesugi, M. Funato, Y. Kawakami, H. Miyake, *Impact of face-to-face annealed sputtered AlN on the optical properties of AlGaN multiple quantum wells*, AIP Advances **9**, 125342 (2019).

[115] K. Shojiki, Y. Hayashi, K. Uesugi, H. Miyake, *Local and anisotropic strain in AlN film on sapphire observed by Raman scattering spectroscopy*, Japanese Journal of Applied Physics **58**, SCCB17 (2019).

[116] S. Tanaka, K. Shojiki, K. Uesugi, Y. Hayashi, H. Miyake, *Quantitative evaluation of strain relaxation in annealed sputter-deposited AlN film*, Journal of Crystal Growth **512**, 16 (2019).

[117] K. Uesugi, Y. Hayashi, K. Shojiki, H. Miyake, *Reduction of threading dislocation density and suppression of cracking in sputter-deposited AlN templates annealed at high temperatures*, Applied Physics Express **12**, 065501 (2019).

[118] K. Uesugi, Y. Hayashi, K. Shojiki, S. Xiao, K. Nagamatsu, H. Yoshida, H. Miyake, *Fabrication of AlN templates on SiC substrates by sputtering-deposition and high-temperature annealing*, Journal of Crystal Growth **510**, 13 (2019).

[119] S. Xiao, N. Jiang, K. Shojiki, K. Uesugi, H. Miyake, *Preparation of high-quality thick AlN layer on nanopatterned sapphire substrates with sputter-deposited annealed AlN film by hydride vapor-phase epitaxy*, Japanese Journal of Applied Physics **58**, SC1003 (2019).

[120] Y. Iba, K. Shojiki, K. Uesugi, S. Xiao, H. Miyake, *MOVPE growth of AlN films on nano-patterned sapphire substrates with annealed sputtered AlN*, Journal of Crystal Growth **532**, 125397 (2020).

[121] S. Kuboya, K. Uesugi, K. Shojiki, Y. Tezen, K. Norimatsu, H. Miyake, *Crystalline quality improvement of face-to-face annealed MOVPE-grown AlN on vicinal sapphire substrate with sputtered nucleation layer*, Journal of Crystal Growth **545**, 125722 (2020).

[122] T. Shirato, Y. Hayashi, K. Uesugi, K. Shojiki, H. Miyake, *High-Temperature Annealing of Sputter-Deposited AlN on (001) Diamond Substrate*, physica status solidi (b) **257**, 1900447 (2019).

[123] S. Teramura, Y. Kawase, Y. Sakuragi, S. Iwayama, M. Iwaya, T. Takeuchi, S. Kamiyama, I. Akasaki, H. Miyake, *High Crystallinity and Highly Relaxed Al 0.60 Ga 0.40 N Films Using Growth Mode Control Fabricated on a Sputtered AlN Template with High-Temperature Annealing*, physica status solidi (a) **217**, 1900868 (2020).

[124] K. Uesugi, K. Shojiki, Y. Tezen, Y. Hayashi, H. Miyake, *Suppression of dislocation-induced spiral hillocks in MOVPE-grown AlGaN on face-to-face annealed sputter-deposited AlN template*, Applied Physics Letters **116**, 062101 (2020).

[125] D. Wang, K. Uesugi, S. Xiao, K. Norimatsu, H. Miyake, *Low dislocation density AlN on sapphire prepared by double sputtering and annealing*, Applied Physics Express **13**, 095501 (2020).

[126] Y. Itokazu, S. Kuwaba, M. Jo, N. Kamata, H. Hirayama, *Influence of the nucleation conditions on the quality of AlN layers with high-temperature annealing and regrowth processes*, Japanese Journal of Applied Physics **58**, SC1056 (2019).

[127] Y. Mogami, S. Motegi, A. Osawa, K. Osaki, Y. Tanioka, A. Maeoka, M. Jo, N. Maeda, H. Yaguchi, H. Hirayama, *Evolution of morphology and crystalline quality of DC-sputtered AlN films with high-temperature annealing*, Japanese Journal of Applied Physics **58**, SC1029 (2019).

[128] M. X. Wang, F. J. Xu, N. Xie, Y. H. Sun, B. Y. Liu, W. K. Ge, X. N. Kang, Z. X. Qin, X. L. Yang, X. Q. Wang, B. Shen, *High-temperature annealing induced evolution of strain in AlN epitaxial films grown on sapphire substrates*, Applied Physics Letters **114**, 112105 (2019).

[129] S. Washiyama, Y. Guan, S. Mita, R. Collazo, Z. Sitar, *Recovery kinetics in high temperature annealed AlN heteroepitaxial films*, Journal of Applied Physics **127**, 115301 (2020).

[130] C. He, W. Zhao, H. Wu, N. Liu, S. Zhang, J. Li, C. Jia, K. Zhang, L. He, Z. Chen, B. Shen, *Fast growth of crack-free thick AlN film on sputtered AlN/sapphire by introducing high-density nano-voids*, Journal of Physics D: Applied Physics **53**, 405303 (2020).

[131] J. E. Ayers, T. Kujofsa, P. Rago, J. Raphael, *Heteroepitaxy of Semiconductors*, CRC Press, 2016.

[132] A. Dadgar, M. Weyers, *Nitride Semiconductors*, in *Metalorganic Vapor Phase Epitaxy (MOVPE): Growth, Materials Properties, and Applications*, pages 109–147, Wiley, 2019.

[133] N. Susilo, E. Ziffer, S. Hagedorn, L. Cancellara, C. Netzel, N. L. Ploch, S. Wu, J. Rass, S. Walde, L. Sulmoni, M. Guttmann, T. Wernicke, M. Albrecht, M. Weyers, M. Kneissl, *Improved performance of UVC-LEDs by combination of high-temperature annealing and epitaxially laterally overgrown AlN/sapphire*, Photonics Research **8**, 589 (2020).

[134] S. Walde, S. Hagedorn, P.-M. Coulon, A. Mogilatenko, C. Netzel, J. Weinrich, N. Susilo, E. Ziffer, L. Matiwe, C. Hartmann, G. Kusch, A. Alasmari, G. Naresh-Kumar, C. Trager-Cowan, T. Wernicke, T. Straubinger, M. Bickermann, R. Martin, P. Shields, M. Kneissl, M. Weyers, *AlN overgrowth of nano-pillar-patterned sapphire with different offcut angle by metalorganic vapor phase epitaxy*, Journal of Crystal Growth **531**, 125343 (2020).

[135] S. Hagedorn, S. Walde, N. Susilo, C. Netzel, N. Tillner, R.-S. Unger, P. Manley, E. Ziffer, T. Wernicke, C. Becker, H.-J. Lugauer, M. Kneissl, M. Weyers, *Improving AlN Crystal Quality and Strain Management on Nanopatterned Sapphire Substrates by High-Temperature Annealing for UVC Light-Emitting Diodes*, physica status solidi (a) **217**, 1900796 (2020).

[136] H. H. Solak, C. Dais, F. Clube, *Displacement Talbot lithography: a new method for high-resolution patterning of large areas*, Optics Express **19**, 10686 (2011).

[137] P.-M. Coulon, B. Damilano, B. Alloing, P. Chausse, S. Walde, J. Enslin, R. Armstrong, S. Vézian, S. Hagedorn, T. Wernicke, J. Massies, J. Zúñiga-Pérez, M. Weyers, M. Kneissl, P. A. Shields, *Displacement Talbot lithography for nano-engineering of III-nitride materials*, Microsystems & Nanoengineering **5**, 1 (2019).

[138] G. Strassburger, A. Dadgar, A. Krost, *Patent 1036179*, DPMA München, Germany (2003).

[139] *The tension of metallic films deposited by electrolysis*, Proceedings of the Royal Society of London. Series A, Containing Papers of a Mathematical and Physical Character **82**, 172 (1909).

[140] F. Brunner, A. Knauer, T. Schenk, M. Weyers, J.-T. Zettler, *Quantitative analysis of in situ wafer bowing measurements for III-nitride growth on sapphire*, Journal of Crystal Growth **310**, 2432 (2008).

[141] S. Hagedorn, *Hydrid-Gasphasenepitaxie zur Herstellung von Aluminiumgalliumnitrid*, Dissertation, Technische Universität Berlin (2015).

[142] S. R. Lee, A. M. West, A. A. Allerman, K. E. Waldrip, D. M. Follstaedt, P. P. Provencio, D. D. Koleske, C. R. Abernathy, *Effect of threading dislocations on the Bragg peakwidths of GaN, AlGaN, and AlN heterolayers*, Applied Physics Letters **86**, 241904 (2005).

[143] P. F. Fewster, *X-Ray Scattering from Semiconductors*, Published by imperial college press and distributed by world scientific publishing co., 2003.

[144] M. A. Moram, M. E. Vickers, *X-ray diffraction of III-nitrides*, Reports on Progress in Physics **72**, 036502 (2009).

[145] C. Dunn, E. Kogh, *Comparison of dislocation densities of primary and secondary recrystallization grains of Si-Fe*, Acta Metallurgica **5**, 548 (1957).

[146] B. N. Pantha, R. Dahal, M. L. Nakarmi, N. Nepal, J. Li, J. Y. Lin, H. X. Jiang, Q. S. Paduano, D. Weyburne, *Correlation between optoelectronic and structural properties and epilayer thickness of AlN*, Applied Physics Letters **90**, 241101 (2007).

[147] P. Gay, P. Hirsch, A. Kelly, *The estimation of dislocation densities in metals from X-ray data*, Acta Metallurgica **1**, 315 (1953).

[148] M. Imura, K. Nakano, N. Fujimoto, N. Okada, K. Balakrishnan, M. Iwaya, S. Kamiyama, H. Amano, I. Akasaki, T. Noro, T. Takagi, A. Bandoh, *Dislocations in AlN Epilayers Grown on Sapphire Substrate by High-Temperature Metal-Organic Vapor Phase Epitaxy*, Japanese Journal of Applied Physics **46**, 1458 (2007).

[149] D. Gerlich, S. Dole, G. Slack, *Elastic properties of aluminum nitride*, Journal of Physics and Chemistry of Solids **47**, 437 (1986).

[150] A. F. Wright, *Elastic properties of zinc-blende and wurtzite AlN, GaN, and InN*, Journal of Applied Physics **82**, 2833 (1997).

[151] G. Haugstad, *Atomic Force Microscopy*, John Wiley & Sons, Inc., 2012.

[152] N. Erdman, D. C. Bell, R. Reichelt, *Scanning Electron Microscopy*, in *Springer Handbook of Microscopy*, pages 229–318, Springer International Publishing, 2019.

[153] C. Trager-Cowan, A. Alasmari, W. Avis, J. Bruckbauer, P. R. Edwards, B. Hourahine, S. Kraeusel, G. Kusch, R. Johnston, G. Naresh-Kumar, R. W. Martin, M. Nouf-Allehiani, E. Pascal, L. Spasevski, D. Thomson, S. Vespucci, P. J. Parbrook, M. D. Smith, J. Enslin, F. Mehnke, M. Kneissl, C. Kuhn, T. Wernicke, S. Hagedorn, A. Knauer, V. Kueller, S. Walde, M. Weyers, P.-M. Coulon, P. A. Shields, Y. Zhang, L. Jiu, Y. Gong, R. M. Smith, T. Wang, A. Winkelmann, *Scanning electron microscopy as a flexible technique for investigating the properties of UV-emitting nitride semiconductor thin films*, Photonics Research **7**, B73 (2019).

[154] C. Trager-Cowan, A. Alasmari, W. Avis, J. Bruckbauer, P. R. Edwards, B. Hourahine, S. Kraeusel, G. Kusch, B. M. Jablon, R. Johnston, R. W. Martin, R. Mcdermott, G. Naresh-Kumar, M. Nouf-Allehiani, E. Pascal, D. Thomson, S. Vespucci, K. Mingard, P. J. Parbrook, M. D. Smith, J. Enslin, F. Mehnke, M. Kneissl, C. Kuhn, T. Wernicke, A. Knauer, S. Hagedorn, S. Walde, M. Weyers, P.-M. Coulon, P. A. Shields, Y. Zhang, L. Jiu, Y. Gong, R. M. Smith, T. Wang,

A. Winkelmann, *Advances in electron channelling contrast imaging and electron backscatter diffraction for imaging and analysis of structural defects in the scanning electron microscope*, IOP Conference Series: Materials Science and Engineering **891**, 012023 (2020).

[155] C. Trager-Cowan, A. Alasmari, W. Avis, J. Bruckbauer, P. R. Edwards, G. Ferenczi, B. Hourahine, A. Kotzai, S. Kraeusel, G. Kusch, R. W. Martin, R. McDermott, G. Naresh-Kumar, M. Nouf-Allehiani, E. Pascal, D. Thomson, S. Vespucci, M. D. Smith, P. J. Parbrook, J. Enslin, F. Mehnke, C. Kuhn, T. Wernicke, M. Kneissl, S. Hagedorn, A. Knauer, S. Walde, M. Weyers, P.-M. Coulon, P. A. Shields, J. Bai, Y. Gong, L. Jiu, Y. Zhang, R. M. Smith, T. Wang, A. Winkelmann, *Structural and luminescence imaging and characterisation of semiconductors in the scanning electron microscope*, Semiconductor Science and Technology **35**, 054001 (2020).

[156] G. Naresh-Kumar, B. Hourahine, P. R. Edwards, A. P. Day, A. Winkelmann, A. J. Wilkinson, P. J. Parbrook, G. England, C. Trager-Cowan, *Rapid Nondestructive Analysis of Threading Dislocations in Wurtzite Materials Using the Scanning Electron Microscope*, Physical Review Letters **108**, 135503 (2012).

[157] G. Naresh-Kumar, D. Thomson, M. Nouf-Allehiani, J. Bruckbauer, P. Edwards, B. Hourahine, R. Martin, C. Trager-Cowan, *Reprint of: Electron channelling contrast imaging for III-nitride thin film structures*, Materials Science in Semiconductor Processing **55**, 19 (2016).

[158] D. B. Williams, C. B. Carter, *Transmission Electron Microscopy*, Springer US, 2009.

[159] L. O. Nyakiti, J. Chaudhuri, E. A. Kenik, P. Lu, J. H. Edgar, *Defect selective etching of thick ALN layers grown on 6H-SiC seeds - A transmission electron microscopy study*, MRS Proceedings **1040**, Q11 (2007).

[160] J. Weyher, *Characterization of wide-band-gap semiconductors (GaN, SiC) by defect-selective etching and complementary methods*, Superlattices and Microstructures **40**, 279 (2006).

[161] P. van der Heide, *Secondary Ion Mass Spectrometry*, John Wiley & Sons, Inc., 2014.

[162] P. Manley, S. Walde, S. Hagedorn, M. Hammerschmidt, S. Burger, C. Becker, *Nanopatterned sapphire substrates in deep-UV LEDs: is there an optical benefit?*, Optics Express **28**, 3619 (2020).

[163] V. Kostylev, *Scattering Fundamentals*, in *Bistatic Radar*, pages 193–223, John Wiley & Sons, Ltd, 2007.

[164] J. Pastrňák, L. Roskovcová, *Refraction Index Measurements on AlN Single Crystals*, physica status solidi (b) **14**, K5 (1966).

[165] *Handbook of optical materials*, Choice Reviews Online **40**, 40 (2003).

[166] J. Pomplun, S. Burger, L. Zschiedrich, F. Schmidt, *Adaptive finite element method for simulation of optical nano structures*, physica status solidi (b) **244**, 3419 (2007).

[167] S. Burger, L. Zschiedrich, J. Pomplun, F. Schmidt, B. Bodermann, *Fast simulation method for parameter reconstruction in optical metrology*, Metrology, Inspection, and Process Control for Microlithography XXVII **8681**, 868119 (2013).

[168] H. J. Monkhorst, J. D. Pack, *Special points for Brillouin-zone integrations*, Physical Review B **13**, 5188 (1976).

[169] V. Liu, S. Fan, *S4 : A free electromagnetic solver for layered periodic structures*, Computer Physics Communications **183**, 2233 (2012).

[170] J. Eisenlohr, N. Tucher, O. Höhn, H. Hauser, M. Peters, P. Kiefel, J. C. Goldschmidt, B. Bläsi, *Matrix formalism for light propagation and absorption in thick textured optical sheets*, Optics Express **23**, A502 (2015).

[171] R. Santbergen, T. Meguro, T. Suezaki, G. Koizumi, K. Yamamoto, M. Zeman, *GenPro4 Optical Model for Solar Cell Simulation and Its Application to Multijunction Solar Cells*, IEEE Journal of Photovoltaics **7**, 919 (2017).

[172] M. Hammerschmidt, *Optical simulation of complex nanostructured solar cells with a reduced basis method*, Dissertation, Freie Universität Berlin (2016).

[173] S. Walde, S. Hagedorn, M. Weyers, *Impact of intermediate high temperature annealing on the properties of AlN/sapphire templates grown by metalorganic vapor phase epitaxy*, Japanese Journal of Applied Physics **58**, SC1002 (2019).

[174] M. Liu, B. Evans, *Dislocation recovery kinetics in single-crystal calcite*, Journal of Geophysical Research: Solid Earth **102**, 24801 (1997).

[175] I. Bryan, Z. Bryan, S. Mita, A. Rice, J. Tweedie, R. Collazo, Z. Sitar, *Surface kinetics in AlN growth: A universal model for the control of surface morphology in III-nitrides*, Journal of Crystal Growth **438**, 81 (2016).

[176] K. Bellmann, U. W. Pohl, C. Kuhn, T. Wernicke, M. Kneissl, *Controlling the morphology transition between step-flow growth and step-bunching growth*, Journal of Crystal Growth **478**, 187 (2017).

[177] L. Zhao, K. Yang, Y. Ai, L. Zhang, X. Niu, H. Lv, Y. Zhang, *Crystal quality improvement of sputtered AlN film on sapphire substrate by high-temperature annealing*, Journal of Materials Science: Materials in Electronics **29**, 13766 (2018).

[178] Y. Wu, A. Hanlon, J. F. Kaeding, R. Sharma, P. T. Fini, S. Nakamura, J. S. Speck, *Effect of nitridation on polarity, microstructure, and morphology of AlN films*, Applied Physics Letters **84**, 912 (2004).

[179] S. Mathis, A. Romanov, L. Chen, G. Beltz, W. Pompe, J. Speck, *Modeling of threading dislocation reduction in growing GaN layers*, Journal of Crystal Growth **231**, 371 (2001).

[180] S. Hagedorn, S. Walde, A. Mogilatenko, M. Weyers, L. Cancellara, M. Albrecht, D. Jaeger, *Stabilization of sputtered AlN/sapphire templates during high temperature annealing*, Journal of Crystal Growth **512**, 142 (2019).

[181] S. Bandyopadhyay, G. Rixecker, F. Aldinger, S. Pal, K. Mukherjee, H. S. Maiti, *Effect of Reaction Parameters on γ-AlON Formation from Al2O3 and AlN*, Journal of the American Ceramic Society **85**, 1010 (2004).

[182] Q. Yan, A. Janotti, M. Scheffler, C. G. V. de Walle, *Origins of optical absorption and emission lines in AlN*, Applied Physics Letters **105**, 111104 (2014).

[183] M. Bickermann, B. M. Epelbaum, O. Filip, P. Heimann, S. Nagata, A. Winnacker, *Point defect content and optical transitions in bulk aluminum nitride crystals*, physica status solidi (b) **246**, 1181 (2009).

[184] B. Sheldon, A. Bhandari, A. Bower, S. Raghavan, X. Weng, J. Redwing, *Steady-state tensile stresses during the growth of polycrystalline films*, Acta Materialia **55**, 4973 (2007).

[185] A. Knauer, A. Mogilatenko, J. Weinrich, S. Hagedorn, S. Walde, T. Kolbe, L. Cancellara, M. Weyers, *The Impact of AlN Templates on Strain Relaxation Mechanisms during the MOVPE Growth of UVB-LED Structures*, Crystal Research and Technology **55**, 1900215 (2020).

[186] C.-Y. Huang, S. Walde, C.-L. Tsai, C. Netzel, H.-H. Liu, S. Hagedorn, Y.-R. Wu, Y.-K. Fu, M. Weyers, *Overcoming the excessive compressive strain in AlGaN epitaxy by introducing high Si-doping in AlN templates*, Japanese Journal of Applied Physics **59**, 070904 (2020).

[187] Y. K. Ooi, J. Zhang, *Light Extraction Efficiency Analysis of Flip-Chip Ultraviolet Light-Emitting Diodes With Patterned Sapphire Substrate*, IEEE Photonics Journal **10**, 1 (2018).

[188] C. Reich, M. Guttmann, M. Feneberg, T. Wernicke, F. Mehnke, C. Kuhn, J. Rass, M. Lapeyrade, S. Einfeldt, A. Knauer, V. Kueller, M. Weyers, R. Goldhahn, M. Kneissl, *Strongly transverse-electric-polarized emission from deep ultraviolet AlGaN quantum well light emitting diodes*, Applied Physics Letters **107**, 142101 (2015).

[189] M. A. Khan, N. Maeda, M. Jo, Y. Akamatsu, R. Tanabe, Y. Yamada, H. Hirayama, *13 mW operation of a 295–310 nm AlGaN UV-B LED with a p-AlGaN transparent contact layer for real world applications*, Journal of Materials Chemistry C **7**, 143 (2019).

[190] T. Takano, T. Mino, J. Sakai, N. Noguchi, K. Tsubaki, H. Hirayama, *Deep-ultraviolet light-emitting diodes with external quantum efficiency higher than 20% at 275 nm achieved by improving light-extraction efficiency*, Applied Physics Express **10**, 031002 (2017).

[191] C. Pernot, M. Kim, S. Fukahori, T. Inazu, T. Fujita, Y. Nagasawa, A. Hirano, M. Ippommatsu, M. Iwaya, S. Kamiyama, I. Akasaki, H. Amano, *Improved Efficiency of 255–280 nm AlGaN-Based Light-Emitting Diodes*, Applied Physics Express **3**, 061004 (2010).

[192] A. Knauer, U. Zeimer, V. Kueller, M. Weyers, *MOVPE growth of AlxGa1-xN with $x \sim 0.5$ on epitaxial laterally overgrown AlN/sapphire templates for UV-LEDs*, physica status solidi (c) **11**, 377 (2014).

[193] A. Knauer, A. Mogilatenko, S. Hagedorn, J. Enslin, T. Wernicke, M. Kneissl, M. Weyers, *Correlation of sapphire off-cut and reduction of defect density in MOVPE grown AlN*, physica status solidi (b) **253**, 809 (2016).

[194] S. Tomiya, K. Funato, T. Asatsuma, T. Hino, S. Kijima, T. Asano, M. Ikeda, *Dependence of crystallographic tilt and defect distribution on mask material in epitaxial lateral overgrown GaN layers*, Applied Physics Letters **77**, 636 (2000).

[195] T. M. Katona, P. Cantu, S. Keller, Y. Wu, J. S. Speck, S. P. DenBaars, *Maskless lateral epitaxial overgrowth of high-aluminum-content AlxGa1-xN*, Applied Physics Letters **84**, 5025 (2004).

[196] Y. Varshni, *Temperature dependence of the energy gap in semiconductors*, Physica **34**, 149 (1967).

[197] U. Zeimer, V. Kueller, A. Knauer, A. Mogilatenko, M. Weyers, M. Kneissl, *High quality AlGaN grown on ELO AlN/sapphire templates*, Journal of Crystal Growth **377**, 32 (2013).

[198] C. B. Soh, S. J. Chua, S. Tripathy, W. Liu, D. Z. Chi, *The influence of V defects on luminescence properties of AlInGaN quaternary alloys*, Journal of Physics: Condensed Matter **17**, 729 (2005).

[199] J. E. Northrup, L. T. Romano, J. Neugebauer, *Surface energetics, pit formation, and chemical ordering in InGaN alloys*, Applied Physics Letters **74**, 2319 (1999).

[200] W. Liu, C. Soh, P. Chen, S. Chua, *Characterization of AlInGaN quaternary epilayers grown by metal organic chemical vapor deposition*, Journal of Crystal Growth **268**, 509 (2004).

[201] I.-H. Kim, H.-S. Park, Y.-J. Park, T. Kim, *Formation of V-shaped pits in InGaN/GaN multiquantum wells and bulk InGaN films*, Applied Physics Letters **73**, 1634 (1998).

[202] S. Walde, M. Brendel, U. Zeimer, F. Brunner, S. Hagedorn, M. Weyers, *Impact of open-core threading dislocations on the performance of AlGaN metal-*

semiconductor-metal photodetectors, Journal of Applied Physics **123**, 161551 (2018).

[203] K. Kojima, Y. Nagasawa, A. Hirano, M. Ippommatsu, Y. Honda, H. Amano, I. Akasaki, S. F. Chichibu, *Carrier localization structure combined with current micropaths in AlGaN quantum wells grown on an AlN template with macrosteps*, Applied Physics Letters **114**, 011102 (2019).

[204] K. Nomura, S. Hanagata, A. Kunisaki, R. Togashi, H. Murakami, Y. Kumagai, A. Koukitu, *High-Temperature Heat-Treatment ofc-,a-,r-, andm-Plane Sapphire Substrates in Mixed Gases of H2and N2*, Japanese Journal of Applied Physics **52**, 08JB10 (2013).

[205] C.-Y. Huang, K.-S. Chang, C.-Y. Huang, Y.-H. Lin, W.-C. Peng, H.-W. Yen, R.-M. Lin, H.-C. Kuo, *The origin and mitigation of volcano-like morphologies in micron-thick AlGaN/AlN heteroepitaxy*, Applied Physics Letters **111**, 072110 (2017).

[206] N. Maeda, H. Hirayama, *Realization of high-efficiency deep-UV LEDs using transparent p-AlGaN contact layer*, physica status solidi (c) **10**, 1521 (2013).

[207] A. G. Mathewson, H. P. Myers, *Absolute Values of the Optical Constants of Some Pure Metals*, Physica Scripta **4**, 291 (1971).

[208] D. Brunner, H. Angerer, E. Bustarret, F. Freudenberg, R. Höpler, R. Dimitrov, O. Ambacher, M. Stutzmann, *Optical constants of epitaxial AlGaN films and their temperature dependence*, Journal of Applied Physics **82**, 5090 (1997).

List of Publications

Parts of this work have been published in the following peer-reviewed articles and conference contributions:

Peer-reviewed articles

- S. Hagedorn, S. Walde, A. Knauer, N. Susilo, D. Pacak, L. Cancellara, C. Netzel, A. Mogilatenko, C. Hartmann, T. Wernicke, M. Kneissl, and M. Weyers, *Status and Prospects of AlN Templates on Sapphire for Ultraviolet Light-Emitting Diodes*, physica status solidi (a) **217**, 1901022 (2020).

- C. Trager-Cowan, A. Alasmari, W. Avis, J. Bruckbauer, P. R. Edwards, B. Hourahine, S. Kraeusel, G. Kusch, B. M. Jablon, R. Johnston, R. W. Martin, R. Mcdermott, G. Naresh-Kumar, M. Nouf-Allehiani, E. Pascal, D. Thomson, S. Vespucci, K. Mingard, P. J. Parbrook, M. D. Smith, J. Enslin, F. Mehnke, M. Kneissl, C. Kuhn, T. Wernicke, A. Knauer, S. Hagedorn, S. Walde, M. Weyers, P.-M. Coulon, P. A. Shields, Y. Zhang, L. Jiu, Y. Gong, R. M. Smith, T. Wang, and A. Winkelmann, *Advances in electron channelling contrast imaging and electron backscatter diffraction for imaging and analysis of structural defects in the scanning electron microscope*, IOP Conference Series: Materials Science and Engineering **891**, 012023 (2020).

- C.-Y. Huang, S. Walde, C.-L. Tasi, C. Netzel, H.-H. Liu, S. Hagedorn, Y.-R. Wuuh, Y.-K. Fu, and M. Weyers, *Overcoming the excessive compressive strain in AlGaN epitaxy by introducing a high Si-doping in AlN templates*, Japanese Journal of Applied Physics **59**, 070904 (2020).

- A. Knauer, A. Mogilatenko, J. Weinrich, S. Hagedorn, S. Walde, T. Kolbe, L. Cancellara, and M. Weyers, *The Impact of AlN Templates on Strain Relaxation Mechanisms during the MOVPE Growth of UVB-LED Structures*, Crystal Research and Technology **55**, 1900215 (2020).

- N. Susilo, E. Ziffer, S. Hagedorn, L. Cancellara, C. Netzel, N. Lobo-Ploch, S. Wu, J. Rass, S. Walde, L. Sulmoni, M. Guttmann, T. Wernicke, M. Albrecht, M. Weyers, and M. Kneissl, *Improved performance of UVC-LEDs by combination of high-temperature annealing and epitaxially laterally overgrown AlN/sapphire*, Photonics Research **8**, 589 (2020).

- S. Hagedorn, S. Walde, N. Susilo, C. Netzel, N. Tillner, R.-S. Unger, P. Manley, E. Ziffer, T. Wernicke, C. Becker, H.-J. Lugauer, M. Kneissl, and M. Weyers, *Improving AlN Crystal Quality and Strain Management on Nanopatterned Sapphire Substrates by High-Temperature Annealing for UVC Light-Emitting Diodes*, physica status solidi (a) **217**, 1900796 (2020).

- C. Trager-Cowan, A. Alasmari, W. Avis, J. Bruckbauer, P. R. Edwards, G. Ferenczi, B. Hourahine, A. Kotzai, S. Kraeusel, G. Kusch, R. W. Martin, R. Mcdermott, G. Naresh-Kumar, M. Nouf-Allehiani, E. Pascal, D. Thomson, S. Vespucci, M. D. Smith, P. J. Parbrook, J. Enslin, F. Mehnke, C. Kuhn, T. Wernicke, M. Kneissl, S. Hagedorn, A. Knauer, S. Walde, M. Weyers, P.-M. Coulon, P. A. Shields, J. Bai, Y. Gong, L. Jiu, R. M. Smith, T. Wang, and A. Winkelmann, *Structural and luminescence imaging and characterisation of semiconductors in the scanning electron microscope*, Semiconductor Science and Technology **35**, 054001 (2020).

- P. Manley, S. Walde, S. Hagedorn, M. Hammerschmidt, S. Burger, and C. Becker, *Nanopatterned Sapphire Substrates: Is there an Optical Benefit?*, Optics Express **28**, 3619 (2020).

- S. Walde, S. Hagedorn, P.-M. Coulon, A. Mogilatenko, C. Netzel, J. Weinrich, N. Susilo, E. Ziffer, L. Matiwe, C. Hartmann, G. Kusch, A. Alasmari, G. Naresh-Kumar, C. Trager-Cowan, T. Wernicke, T. Straubinger, M. Bickermann, R. W. Martin, P. A. Shields, M. Kneissl, and M. Weyers, *AlN overgrowth of nano-pillar-patterned sapphire with different offcut angle by metalorganic vapor phase epitaxy*, Journal of Crystal Growth **531**, 125343 (2020).

- P.-M. Coulon, B. Damilano, B. Alloing, P. Chausse, S. Walde, J. Enslin, R. Armstrong, S. Vezian, S. Hagedorn, T. Wernicke, J. Massies, J. Zuniga-Perez, M. Weyers, M. Kneissl, and P.A. Shields, *Displacement Talbot Lithography for nano-engineering of III-nitride materials*, Microsystems and Nanoengineering **5**, 52 (2019).

- C. Trager-Cowan, A. Alasmari, W. Avis, J. Bruckbauer, P. R. Edwards, B. Hourahine, S. Kraeusel, G. Kusch, R. Johnston, G. Naresh-Kumar, R. W. Martin, M. Nouf-Allehiani, E. Pascal, L. Spasevski, D. Thomson, S. Vespucci, P. J. Parbrook, M. D. Smith, J. Enslin, F. Mehnke, M. Kneissl, C. Kuhn, T. Wernicke, S. Hagedorn, A. Knauer, V. Kueller, S. Walde, M. Weyers, P.-M. Coulon, P. A. Shields, Y. Zhang, L. Jiu, Y. Gong, R. M. Smith, T. Wang, and A. Winkelmann, *Scanning electron microscope as a flexible tool for investigating the properties of UV-emitting nitride semiconductors thin films*, Photonics Research **7**, B73 (2019).

- S. Walde, S. Hagedorn, and M. Weyers, *Impact of intermediate high temperature annealing on the properties of AlN/sapphire templates grown by metalorganic vapor phase epitaxy*, Japanese Journal of Applied Physics **58**, SC1002 (2019).
- S. Hagedorn, S. Walde, A. Mogilatenko, M. Weyers, L. Cancellara, M. Albrecht, and D. Jäger, *Stabilization of sputtered AlN/sapphire templates during high temperature annealing*, Journal of Crystal Growth **512**, 142 (2019).

Conference contributions

- S. Walde, S. Hagedorn, N. Tillner, H.-J. Lugauer, and M. Weyers, *AlN overgrowth of nano-hole patterned sapphire by metalorganic vapor phase epitaxy*, Contributed talk, Workshop der Deutschen Gesellschaft für Kristallwachstum und Kristallzüchtung 2019, Dresden, Germany.
- S. Walde, S. Hagedorn, A. Knauer, A. Mogilatenko, N. Susilo, B. Belde, L. Sulmoni, T. Wernicke, M. Kneissl, and M. Weyers, *AlN/sapphire templates for UV LEDs: Different approaches and their impact on device performance*, Invited talk, International workshop on UV materials and devices 2018, Kunming, China.
- S. Walde, S. Hagedorn, N. Susilo, A. Knauer, T. Wernicke, M. Kneissl, and M. Weyers, *Impact of intermediate high temperature annealing on the properties of MOVPE grown AlN/sapphire templates*, Contributed talk, International Workshop on Nitride Semiconductors 2018, Kanazawa, Japan.
- S. Walde, S. Hagedorn, P.-M. Coulon, N. Susilo, G. Kusch, N. Suss, L. Matiwe, C. Hartmann, G. Naresh-Kumar, C. Trager-Cowan, A. Knauer, T. Wernicke, M. Bickermann, R. W. Martin, P. A. Shields, M. Kneissl, and M. Weyers, *MOVPE grown AlN on nano-patterned sapphire substrates with different offcut angle*, Contributed talk, International Symposium on Growth of III-Nitrides 2018, Warsaw, Poland.
- S. Walde, S. Hagedorn, U. Zeimer, A. Knauer, and M. Weyers, *AlN growth on nano-patterned sapphire substrates*, Contributed talk, Workshop der Deutschen Gesellschaft für Kristallwachstum und Kristallzüchtung 2017, Freiburg, Germany.

List of Figures

Acknowledgement

Before closing this work, I want to cordially thank everyone, who contributed directly or indirectly to this work:

I thank the director of the Ferdinand-Braun-Institut Prof. Dr. Günther Tränkle for providing an excellent scientific environment and the necessary resources during my PhD. Furthermore, I am thankful for his constant interest in my research work, which led to motivating discussions on a regular basis.

I thank Prof. Dr. Michael Kneissl for supervising my work as the first referee at the Technical University of Berlin and for the opportunity to present the progress of my research yearly in the seminar of his group.

I thank Prof. Dr. Markus Weyers for supervising my work in his department Materials Technology at the Ferdinand-Braun-Institut. I want to especially thank him for the always very short communication way in all kinds of matters and the support in connecting with other scientists all around the world during international conferences and extended research stays.

I thank Prof. Dr. Ferdinand Scholz for evaluating the PhD thesis as the external referee.

I thank Dr. Sylvia Hagedorn for the supervision of my work on a daily basis. Her scientific expertise combined with the open and friendly style of working together created the perfect environment for me to progress in my scientific competencies. Additionally, I thank all members of our small group related to AlN on sapphire, Torsten Petzke, Daniel Pacak, Taimoor Khan and Franz Kaiser.

I thank all collaboration partners for their interest in the joint work, namely Dr. Pierre-Marie Coulon, Dr. Philip Shields, Dr. Phillip Manley, Prof. Dr. Christiane Becker, Dr. Carlo Barth, Dr. Chia-Yen Huang, Dr. Yi-Keng Fu, Prof. Dr. Yuh-Renn Wu, Leonardo Cancellara, Dr. Martin Albrecht, Dr. Carsten Hartmann, Lucinda Matiwe, Dr. Gunnar Kusch, Dr. Carol-Trager Cowan, Nadine Tillner, Dr. Hans-Jürgen Lugauer,

Dr. Norman Susilo, Dr. Martin Guttmann, Dr. Tim Wernicke, Dr. Ralph-Stephan Unger, Dr. Arne Knauer, Dr. Tim Kolbe, Dr. Carsten Netzel, Helen Lawrenz, Dr. Anna Mogilatenko, Jonas Weinrich, Dr. Eberhard Richter. The collaborations resulted in many different kinds of contributions to this work, e.g. measurements, UV LED growth and simulations, but also scientific discussions.

I thank all colleagues of the Ferdinand-Braun-Institut, especially the Materials Technology Department, the GaN Optoelectronics Department, the Management Department and the Process Technology Department for the support and help regarding numerous issues and for the warm and friendly working atmosphere.

Conclusively, I thank my family and my friends. This project would not have been possible without them. A special thank you goes to my parents Ruth and Wilhelm, who laid the foundation of my academic path in my education.

Innovationen mit Mikrowellen und Licht
Forschungsberichte aus dem Ferdinand-Braun-Institut, Leibniz-Institut für Höchstfrequenztechnik

Herausgeber: Prof. Dr. G. Tränkle

Band 1: **Thorsten Tischler**
Die Perfectly-Matched-Layer-Randbedingung in der Finite-Differenzen-Methode im Frequenzbereich: Implementierung und Einsatzbereiche
ISBN: 3-86537-113-2, 19,00 EUR, 144 Seiten

Band 2: **Friedrich Lenk**
Monolithische GaAs FET- und HBT-Oszillatoren mit verbesserter Transistormodellierung
ISBN: 3-86537-107-8, 19,00 EUR, 140 Seiten

Band 3: **R. Doerner, M. Rudolph (eds.)**
Selected Topics on Microwave Measurements, Noise in Devices and Circuits, and Transistor Modeling
ISBN: 3-86537-328-3, 19,00 EUR, 130 Seiten

Band 4: **Matthias Schott**
Methoden zur Phasenrauschverbesserung von monolithischen Millimeterwellen-Oszillatoren
ISBN: 978-3-86727-774-0, 19,00 EUR, 134 Seiten

Band 5: **Katrin Paschke**
Hochleistungsdiodenlaser hoher spektraler Strahldichte mit geneigtem Bragg-Gitter als Modenfilter (α-DFB-Laser)
ISBN: 978-3-86727-775-7, 19,00 EUR, 128 Seiten

Band 6: **Andre Maaßdorf**
Entwicklung von GaAs-basierten Heterostruktur-Bipolartransistoren (HBTs) für Mikrowellenleistungszellen
ISBN: 978-3-86727-743-3, 23,00 EUR, 154 Seiten

Band 7: **Prodyut Kumar Talukder**
Finite-Difference-Frequency-Domain Simulation of Electrically Large Microwave Structures using PML and Internal Ports
ISBN: 978-3-86955-067-1, 19,00 EUR, 138 Seiten

Band 8: **Ibrahim Khalil**
Intermodulation Distortion in GaN HEMT
ISBN: 978-3-86955-188-3, 23,00 EUR, 158 Seiten

Band 9: **Martin Maiwald**
Halbleiterlaser basierte Mikrosystemlichtquellen für die Raman-Spektroskopie
ISBN: 978-3-86955-184-5, 19,00 EUR, 134 Seiten

Band 10: **Jens Flucke**
Mikrowellen-Schaltverstärker in GaN- und GaAs-Technologie Designgrundlagen und Komponenten
ISBN: 978-3-86955-304-7, 21,00 EUR, 122 Seiten

Cuvillier Verlag
Internationaler wissenschaftlicher Fachverlag

Innovationen mit Mikrowellen und Licht
Forschungsberichte aus dem Ferdinand-Braun-Institut, Leibniz-Institut für Höchstfrequenztechnik

Herausgeber: Prof. Dr. G. Tränkle

Band 11: **Harald Klockenhoff**
Optimiertes Design von Mikrowellen-Leistungstransistoren und Verstärkern im X-Band
ISBN: 978-3-86955-391-7, 26,75 EUR, 130 Seiten

Band 12: **Reza Pazirandeh**
Monolithische GaAs FET- und HBT-Oszillatoren mit verbesserter Transistormodellierung
ISBN: 978-3-86955-107-8, 19,00 EUR, 140 Seiten

Band 13: **Tomas Krämer**
High-Speed InP Heterojunction Bipolar Transistors and Integrated Circuits in Transferred Substrate Technology
ISBN: 978-3-86955-393-1, 21,70 EUR, 140 Seiten

Band 14: **Phuong Thanh Nguyen**
Investigation of spectral characteristics of solitary diode lasers with integrated grating resonator
ISBN: 978-3-86955-651-2, 24,00 EUR, 156 Seiten

Band 15: **Sina Riecke**
Flexible Generation of Picosecond Laser Pulses in the Infrared and Green Spectral Range by Gain-Switching of Semiconductor Lasers
ISBN: 978-3-86955-652-9, 22,60 EUR, 136 Seiten

Band 16: **Christian Hennig**
Hydrid-Gasphasenepitaxie von versetzungsarmen und freistehenden GaN-Schichten
ISBN: 978-3-86955-822-6, 27,00 EUR, 162 Seiten

Band 17: **Tim Wernicke**
Wachstum von nicht- und semipolaren InAlGaN-Heterostrukturen für hocheffiziente Licht-Emitter
ISBN: 978-3-86955-881-3, 23,40 EUR, 138 Seiten

Band 18: **Andreas Wentzel**
Klasse-S Mikrowellen-Leistungsverstärker mit GaN-Transistoren
ISBN: 978-3-86955-897-4, 29,65 EUR, 172 Seiten

Band 19: **Veit Hoffmann**
MOVPE growth and characterization of (In,Ga)N quantum structures for laser diodes emitting at 440 nm
ISBN: 978-3-86955-989-6, 18,00 EUR, 118 Seiten

Band 20: **Ahmad Ibrahim Bawamia**
Improvement of the beam quality of high-power broad area semiconductor diode lasers by means of an external resonator
ISBN: 978-3-95404-065-0, 21,00 EUR, 126 Seiten

Cuvillier Verlag
Internationaler wissenschaftlicher Fachverlag

Innovationen mit Mikrowellen und Licht
Forschungsberichte aus dem Ferdinand-Braun-Institut, Leibniz-Institut für Höchstfrequenztechnik

Herausgeber: Prof. Dr. G. Tränkle

Band 21: **Agnietzka Pietrzak**
Realization of High Power Diode Lasers with Extremely Narrow Vertical Divergence
ISBN: 978-3-95404-066-7, 27,40 EUR, 144 Seiten

Band 22: **Eldad Bahat-Treidel**
GaN-based HEMTs for High Voltage Operation
Design, Technology and Characterization
ISBN: 978-3-95404-094-0, 41,10 EUR, 220 Seiten

Band 23: **Ponky Ivo**
AlGaN/GaN HEMTs Reliability:
Degradation Modes and Anslysis
ISBN: 978-3-95404-259-3, 23,55 EUR, 132 Seiten

Band 24: **Stefan Spießberger**
Compact Semiconductor-Based Laser Sources
with Narrow Linewidth and High Output Power
ISBN: 978-3-95404-261-6, 24,15 EUR, 140 Seiten

Band 25: **Silvio Kühn**
Mikrowellenoszillatoren für die Erzeugung von atmosphärischen Mikroplasmen
ISBN: 978-3-95404-378-1, 21,85 EUR, 112 Seiten

Band 26: **Sven Schwertfeger**
Experimentelle Untersuchung der Modensynchronisation in Multisegment-Laserdioden zur Erzeugung kurzer optischer Pulse bei einer Wellenlänge von 920 nm
ISBN: 978-3-95404-471-9, 29,45 EUR, 150 Seiten

Band 27: **Christoph Matthias Schultz**
Analysis and mitigation of the factors limiting the effiency of high power distributed feedback diode lasers
ISBN: 978-3-95404-521-1, 68,40 EUR, 388 Seiten

Band 28: **Luca Redaelli**
Design and fabrication of GaN-based laser diodes for single-mode and narrow-linewidth applications
ISBN: 978-3-95404-586-0, 29,70 EUR, 176 Seiten

Band 29: **Martin Spreemann**
Resonatorkonzepte für Hochleistungs-Diodenlaser
mit ausgedehnten lateralen Dimensionen
ISBN: 978-3-95404-628-7, 25,15 EUR, 128 Seiten

Cuvillier Verlag
Internationaler wissenschaftlicher Fachverlag

Innovationen mit Mikrowellen und Licht
Forschungsberichte aus dem Ferdinand-Braun-Institut, Leibniz-Institut für Höchstfrequenztechnik

Herausgeber: Prof. Dr. G. Tränkle

Band 30:	**Christian Fiebig** Diodenlaser mit Trapezstruktur und hoher Brillanz für die Realisierung einer Frequenzkonversion auf einer mikro-optischen Bank ISBN: 978-3-95404-690-4, 26,30 EUR, 140 Seiten
Band 31:	**Viola Küller** Versetzungsreduzierte AlN- und AlGaN-Schichten als Basis für UV LEDs ISBN: 978-3-95404-741-3, 34,40 EUR, 164 Seiten
Band 32:	**Daniel Jedrzejczyk** Efficient frequency doubling of near-infrared diode lasers using quasi phase-matched waveguides ISBN: 978-3-95404-958-5, 27,90 EUR, 134 Seiten
Band 33:	**Sylvia Hagedorn** Hybrid-Gasphasenepitaxie zur Herstellung von Aluminiumgalliumnitrid ISBN: 978-3-95404-985-1, 38,00 EUR, 176 Seiten
Band 34:	**Alexander Kravets** Advanced Silicon MMICs for mm-Wave Automotive Radar Front-Ends ISBN: 978-3-95404-986-8, 31,90 EUR, 156 Seiten
Band 35:	**David Feise** Longitudinale Modenfilter für Kantanemitter im roten Spektralbereich ISBN: 978-3-7369-9116-3, 39,20 EUR, 168 Seiten
Band 36:	**Ksenia Nosaeva** Indium phosphide HBT in thermally optimized periphery for applications up to 300GHZ ISBN: 978-3-7369-287-0, 42,00 EUR, 154 Seiten
Band 37:	**Muhammad Maruf Hossain** Signal Generation for Millimeter Wave and THZ Applications in InP-DHBT and InP-on-BiCMOS Technologies ISBN: 978-3-7369-9335-8, 35,60 EUR, 136 Seiten
Band 38:	**Sirinpa Monayakul** Development of Sub-mm Wave Flip-Chip Interconnect ISBN: 978-3-7369-9410-2, 44,00 EUR, 146 Seiten
Band 39:	**Moritz Brendel** Charakterisierung und Optimierung von (Al, Ga) N-basierten UV-Photodetektoren ISBN: 978-3-7369-9465-2, 49,90 EUR, 196 Seiten
Band 40:	**Erdenetsetseg Luvsandamdin** Development of micro-integrated diode lasers for precision quantum optics experiments in space ISBN: 978-3-7369-9479-9, 39,00 EUR, 126 Seiten

Internationaler wissenschaftlicher Fachverlag

Innovationen mit Mikrowellen und Licht
Forschungsberichte aus dem Ferdinand-Braun-Institut, Leibniz-Institut für Höchstfrequenztechnik

Herausgeber: Prof. Dr. G. Tränkle

Band 41:	**Thi Nghiem Vu** Development and analysis of diode laser ns-MOPA systems for high peak power application ISBN: 978-3-7369-9480-5, 38,80 EUR, 138 Seiten
Band 42:	**Christian Bansleben** Differentieller Mikrowellen-Leistungsoszillator für die Realisierung ultrakompakter Plasmaquellen in Matrixanordnung ISBN: 978-3-7369-9530-7, 34,90 EUR, 136 Seiten
Band 43:	**Martin Winterfeldt** Investigation of slow-axis beam quality degradation in high-power broad area diode lasers ISBN: 978-3-7369-9733-2, 39,90 EUR, 158 Seiten
Band 44:	**Jonathan Decker** Investigation of monolithically integrated spectral stabilization in high-brightness broad area diode lasers ISBN: 978-3-7369-9798-1, 49,50 EUR, 174 Seiten
Band 45:	**Andreea Cristina Andrei** Untersuchung und Optimierung robuster und hochlinearer rauscharmer Verstärker in GaN-Technologie ISBN: 978-3-7369-9810-0, 39,90 EUR, 148 Seiten
Band 46:	**Peng Luo** GaN HEMT Modeling Including Trapping Effects Based on Chalmers Model and Pulsed S-Parameter Measurements ISBN: 978-3-7369-9906-0, 48,00 EUR, 160 Seiten
Band 47:	**Jörg Jeschke** Entwicklung von optisch pumpbaren UVC-Lasern auf AlGaN-Basis Chalmers Model and Pulsed S-Parameter Measurements ISBN: 978-3-7369-9918-3, 44,90 EUR, 176 Seiten
Band 48:	**Nikolai Wolff** Wideband GaN Microwave Power Amplifiers with Class-G Supply Modulation ISBN: 978-3-7369-9931-2, 44,90 EUR, 170 Seiten
Band 49:	**Carlo Frevert** Optimization of broad-area GaAs diode lasers for high powers and high efficiences in the temperature range 200-220 K ISBN: 978-3-7369-9944-2, 44,90 EUR, 174 Seiten
Band 50:	**Bassem Arar** GaAs-based components for photonic integrated circuits ISBN: 978-3-7369-9976-3, 43,60 EUR, 152 Seiten

Cuvillier Verlag
Internationaler wissenschaftlicher Fachverlag

Innovationen mit Mikrowellen und Licht
Forschungsberichte aus dem Ferdinand-Braun-Institut, Leibniz-Institut für Höchstfrequenztechnik

Herausgeber: Prof. Dr. G. Tränkle

Band 51: **Mahmoud Tawfieq**
Development and characterisation of a diode laser based tunable high-power MOPA system
ISBN: 978-3-7369-9983-1, 44,90 EUR, 170 Seiten

Band 52: **Sebastian Preis**
Hocheffiziente frequenzagile Mikrowellen-Leistungsverstärker auf Basis von Verbindungshalbleitern und Ferroelektrika
ISBN: 978-3-7369-7004-5, 54,00 EUR, 136 Seiten

Band 53: **Simon Fleischmann**
Materialaspekte der Hydridgasphasenepitaxie von Aluminiumgalliumnitrid
ISBN: 978-3-7369-7029-8, 41,80 EUR, 160 Seiten

Band 54: **Erhan Ersoy**
Optimierung von koplanaren GaN-MMIC-Leistungsverstärkern im X-Band
ISBN: 978-3-7369-7157-8, 39,90 EUR, 154 Seiten

Band 55: **Simon Rauch**
Lateral emission characteristics of high-power broad-area lasers subject to external optical feedback
ISBN: 978-3-7369-7168-5, 29,90 EUR, 112 Seiten

Band 56: **Frank Dittmar**
Untersuchung der Strahlgüte von brillanten Hochleistungs-Trapezlasern für den Wellenlängenbereich bei 808 nm
ISBN: 978-3-7369-7167-7, 44,90 EUR, 176 Seiten

Band 57: **Florian Hühn**
Flexibler Modulator und Digitalverstärker-MMIC für den energieeffizienten Betrieb einer digitalen Sendekette im GHz-Bereich
ISBN: 978-3-7369-7190-5, 29,90 EUR, 162 Seiten

Band 58: **Dimitri Stoppel**
Interconnection development for InP-HBT terahertz circuits
ISBN: 978-3-7369-7204-9, 44,90 EUR, 156 Seiten

Band 59: **Anissa Zeghuzi**
Analysis of Spatio-Temporal Phenomena in High-Brightness Diode Lasers using Numerical Simulations
ISBN: 978-3-7369-7289-6, 49,90 EUR, 176 Seiten

Band 60: **Tasmim Alam**
Spectroscopic Applications of Terahertz Quantum-Cascade Lasers
ISBN: 978-3-7369-7297-1, 39,90 EUR, 132 Seiten

Cuvillier Verlag
Internationaler wissenschaftlicher Fachverlag

Innovationen mit Mikrowellen und Licht
Forschungsberichte aus dem Ferdinand-Braun-Institut, Leibniz-Institut für Höchstfrequenztechnik

Herausgeber: Prof. Dr. G. Tränkle

Band 61: **Max Schiemangk**
Ein Lasersystem für Experimente mit Quantengasen unter Schwerelosigkeit
ISBN: 978-3-7369-7293-3, 49,90 EUR, 166 Seiten

Band 62: **Thorben Kaul**
Epitaxial Design Optimizations for Increased Efficiency in GaAs-Based High Power Diode Lasers
ISBN: 978-3-7369-7396-1, 38,90 EUR, 136 Seiten

Band 63: **Pietro della Casa**
Two Step MOPVE, in-situ etching and buried implantations: applications to the realization of GaAs laser diodes
ISBN: 978-3-7369-7397-8, 72,90 EUR, 250 Seiten

Band 64: **Heike Christopher**
A compact mode-locked diode laser system for high prescision frequency comparison experiments
ISBN: 978-3-7369-7399-2, 59,90 EUR, 206 Seiten

Band 65: **Rimma Zhytnytska**
Design von GaN Transistoren für leistungselektronische Anwendungen
ISBN: 978-3-7369-7413-5, 57,90 EUR, 182 Seiten

Cuvillier Verlag
Internationaler wissenschaftlicher Fachverlag

www.ingramcontent.com/pod-product-compliance
Ingram Content Group UK Ltd.
Pitfield, Milton Keynes, MK11 3LW, UK
UKHW022000190726
13853UKWH00004B/1647

9 783736 974517